CEREBRUM 2016

Cerebrum 2016
Emerging Ideas in Brain Science

Bill Glovin, Editor

DANA
PRESS

New York

Published by Dana Press, a Division of the Charles A. Dana Foundation, Incorporated

Address correspondence to:
Dana Press
505 Fifth Avenue, Sixth Floor
New York, NY 10017

THE
DANA
FOUNDATION

New York, NY 10017
DANA is a federally registered trademark.
Printed in the United States of America
ISBN-13: 978-1-932594-60-7
ISSN: 1524-6205
Book design by Bruce Hanson at EGADS (egadsontheweb.com)
Cover illustration by William Hogan

CONTENTS

vulsions can vary widely. A new study has found that inhibiting an enzyme that is critical in metabolic communication has an antiseizure effect in epileptic mice. These findings, our authors believe, may very well initiate a shift to new therapeutic approaches.

Raymond Dingledine, Ph.D., and Bjønar Hassel, M.D., Ph.D.

Funding for neuroscience has lagged behind cancer and cardiology for decades. But things are starting to change. From the White House's Brain Initiative to the Ice Bucket Challenge for ALS to some recent sizeable gifts to universities, money for brain research appears to be on the rise. But, as our author explains, research and development funding from private and corporate lenders for cognitive neuroscience—an area he tracks—is also vital.

Harry M. Tracy, Ph.D.

While Flint, Michigan, fueled outrage and heightened awareness about the hazards of lead in tap water, the problem has existed in for years in the US and in other countries. Our author, a winner of the MacArthur Foundation "genius" grant for her work in identifying preventable causes of human disease related to environmental exposures, points out that problems extend well beyond lead. Many potentially harmful contaminants have yet to be evaluated, much less regulated.

Ellen K. Silbergeld, Ph.D.

At a conference in April 2016 in Washington, D.C., the World Bank Group, together with the World Health Organization and other partners, kick-started a call to action to governments, international partners, health professionals, and others to find solutions to a rising global mental health problem. Our authors write that mental disorders account for 30 percent of the nonfatal disease burden worldwide and 10 percent of overall disease burden, including death and disability, and that the global cost—estimated to be approximately $2.5 trillion in 2010—is expected to rise to $6 trillion by 2030.

Shekhar Saxena, M.D., and Patricio V. Marquez, Sc.M.

As the first phase of one of the most ambitious projects in the history of neuroscience comes to a close, one early and influential leader and his younger colleague explain its evolution and underpinnings. Its goal is to build a 'network map' that will shed light on the anatomical and functional connectivity within the healthy human brain, as well as to produce a body of data that will facilitate research into brain disorders.

David C. Van Essen, Ph.D., and Matthew F. Glasser, Ph.D.

Ramón y Cajal, a mythic figure in science and recognized as the father of modern anatomy and neurobiology, was largely responsible for the modern conception of the brain. The first to publish on the nervous system, he sought to educate the novice scientist about how he thought science should be done. We asked an accomplished young investigator to take a fresh look at this recently rediscovered classic, first published in 1897.

Michael L. Anderson, Ph.D.

Waterboarding, sleep deprivation, and solitary confinement were some of the tactics outlined and authorized in a series of Bush administration secret legal documents known as the "torture memos," which were made public in 2009. O'Mara's book casts morality aside to examine whether torture produces reliable information. He reviews existing research in psychology and neuroscience to highlight the impact of torture methods on brain function.

Moheb Costandi, M.Sc.

Foreword

By Guy McKhann, M.D.

Guy McKhann, M.D., the scientific advisor to the Dana Foundation, studies the delineation of the neurological outcomes following coronary artery bypass grafting and the elucidation of the mechanism of a form of Guillen Barre Syndrome. He is professor of neurology at the Johns Hopkins University School of Medicine, with a joint appointment in the school's Department of Neuroscience. McKhann attended Harvard University and obtained his doctoral degree from Yale Medical School. After working at the National Institute of Neurological Disorders and Stroke, he took a residency in pediatric neurology at Massachusetts General Hospital. He then moved to Johns Hopkins University Medical Center, where he was the first director of the neurology department and the founding director of the university's Zanvyl Krieger Mind/Brain Institute. McKhann has authored more than 200 publications and was co-editor for many years of the neurology textbook *Diseases of the Nervous System: Clinical Neurobiology.* He and his colleague (and wife), Marilyn Albert, Ph.D., published a book about aging and the brain for the general public, *Keep Your Brain Young.* McKhann has been involved with a number of scientific organizations, including, as president of the American Neurological Association.

THERE ARE SEVERAL WAYS I might write the Foreword to the *Cerebrum Anthology* for 2016. One is to comment on each article—an approach that has been used in some previous Forewords. A second might be to focus on one or two articles as a basis for further discussion. I have chosen not to use either. Rather than look back at the many fine articles published in the past year, I choose to look forward to what is no doubt coming: an advance in genetics that will change almost all aspects of medicine, whether it be neurological diseases, infectious diseases, or cancer. That advance is called "gene editing," a process through which the normal stability of our genetically controlled activities is modified—not over many decades or more as occurs in the process of evolution, but (in evolutionary terms) almost instantaneously.

The advance that has made this possible is called CRISPR-Cas9 (CRISPR stands for "clustered regularly interspaced short palindromic repairs"), a unique technology that enables geneticists and medical researchers to edit parts of the genome by removing, adding, or altering sections of the DNA sequence.

We are talking about a methodology that has been around for only three to four years. It was selected as the breakthrough of the year by *Science* magazine in 2015 and, as judged by the increased number of articles about CRISPR, has become one of the hottest areas in biology and the source of significant buzz throughout the science world.

What is all the fuss about? This technology offers a quick, economical, and simple means to alter the genome (the genetic makeup) of a plant or animal—human beings included. CRISPR will allow us to increase or decrease the activity of a gene, eliminate a mutation, or induce a new one. Applications to diseases of the nervous system are just starting, using mouse models of autism or schizophrenia. Seldom do we encounter a new, easy-to-use technique that has the ability to so substantially enlighten us about the brain. I would put CRISPR in the same category as the remarkable advances that have come with neuroimaging.

This advance comes at a most opportune time in neurological research. I teach an undergraduate course at Johns Hopkins titled Diseases and Disorders of the Nervous System. I don't actually teach most of the course; what I do is con my colleagues into visiting the class to tell the students what is going on in their particular field. It doesn't matter whether the subject is schizophrenia, Alzheimer's disease, head injury, or muscle disease, my colleagues all wind up saying the same thing: understanding the mechanism of the disease and developing approaches to new therapies are dependent on insight into the underlying genetics.

An example of the use of CRISPR was recently outlined by Michael Specter in the January 2, 2017, issue of *The New Yorker*, not a usual source of scientific information. The article, "Rewriting the Code of Life. Through DNA Editing Researchers Hope to Alter the Genetic Destiny of Species and Eliminate Diseases," details how a group at MIT proposes to eliminate Lyme Disease by modifying the white mouse, one of the animals that is part of the chain of infection of the disease that goes through the tick to the human. They propose to do this by either making the female mouse infertile or making the mouse no longer a host to the Lyme agent. They plan to test this approach in a specific environment, Nantucket, which was chosen because there is a very high rate of human infection by Lyme Disease, as well as a large white mouse population.

Obvious ethical considerations surround this technology, particularly its use in humans. The major concern is introducing changes in germ lines; that is, in cells that would carry genetic modifications into future generations. In the US and Britain strong statements have been issued against using CRISPR in people. However, the technology is so easy to apply that other countries, not bound by US mandates, are likely to use it. Future discussions of this topic in *Cerebrum* should include a rigorous conversation about the ethics of CRISPR, in addition to but not to the exclusion of its scientific potential.

ARTICLES

1

The Changing Face of Recreational Drug Use

By Michael H. Baumann, Ph.D.

Michael H. Baumann, Ph.D., is a staff scientist and facility head at the National Institute on Drug Abuse, Intramural Research Program, in Baltimore, MD. Baumann's research focuses on the role of brain dopamine and serotonin systems in mediating the effects of therapeutic and abused stimulant drugs. In 2012, he joined the laboratory of Amy H. Newman, Ph.D., where he established the Designer Drug Research Unit (DDRU). The main goal of the DDRU is to collect, analyze, and disseminate the most up-to-date information about the pharmacology and toxicology of newly emerging designer drugs of abuse, more formally known as new psychoactive substances (NPS). Working with partner organizations such as the Drug Enforcement Administration, the National Drug Early Warning System, and the European Monitoring Centre for Drugs and Drug Addiction, Baumann is kept informed about recent trends in the abuse of NPS. Most recently, his research has characterized the molecular mechanism of action and pharmacological effects for many of the so-called "bath salts" cathinones and their various replacement analogs.

 *Our author writes that recent data from the United Nations Office of Drugs and Crime indicate that 540 different drugs classified as new psychoactive substances (NPS) have been identified worldwide as of 2014, and this number is expected to rise. His article describes the complexity of the NPS problem, what is known about the molecular mechanisms of action, and the pharmacological effects of NPS. It also highlights some of the considerable challenges in dealing with this emerging issue.*

DRUG ABUSE IS A PERSISTENT public health problem in modern society, and a disturbing new trend is the increased recreational use of so-called "designer drugs," "legal highs," or "research chemicals." These drugs, collectively known as "new psychoactive substances" (NPS), are synthetic alternatives to traditional illegal drugs of abuse.[1]

Most NPS are manufactured by Asian laboratories and sold to consumers via the Internet or shipped to locations in Europe, the United States, and elsewhere, to be packaged for retail sale. NPS are usually marketed as nondrug products to minimize legal scrutiny. They are given innocuous names and labeled "not for human consumption."

Compared to traditional drugs of abuse, NPS are cheap, easy to obtain, and not detectable by standard toxicology screens. There are popular examples of NPS that appear to mimic traditionally abused drugs—stimulant-like NPS (e.g., "bath salts"), marijuana-like NPS (e.g., "spice"), and LSD-like NPS (e.g., "N-bombs")—but no controlled clinical laboratory investigations have been carried out to examine the psychoactive effects of these new drugs in humans.

Nevertheless, the adverse side effects of NPS in humans are well-documented in the medical literature, indicating that these substances pose obvious health risks. There is no quality control in the manufacture and packaging of these products. Adverse effects that have been reported include agitation, panic attacks, hallucinations, psychosis, violent behaviors, increased heart rate, elevated body temperature, and seizures—some of these ending in death.[2] In the US, an alarming spike in toxic exposures and fatalities associated with the abuse of synthetic marijuana-like drugs has occurred during the past year, illustrating the severity and scope of the problem.[3]

Information freely available on the Internet has facilitated the current rise in availability and use of NPS.[4] Scientific articles published in online databases (e.g., PubMed) provide step-by-step recipes for the syntheses of psychoactive compounds, most of which were originally developed as potential medicines or research tools. Such synthetic schemes can be exploited by skilled and corrupt individuals, or businesses, to produce bulk quantities of NPS that are marketed and sold to consumers via the worldwide web. In most countries, adolescents cannot legally buy cigarettes or alcohol, but they can easily purchase powerful psychoactive drugs from websites. Online user forums contain detailed "trip reports" that describe the doses, preferred routes of administration, and expected psychological effects for various NPS, so that users can fine-tune their drug-taking experiences.[5]

As governments have passed legislation to ban specific problematic NPS, chemists involved with the manufacture of these substances have consulted the scientific or patent literatures and quickly created novel "replacement" drugs to stay one step ahead of law enforcement. The sheer number of new drugs now is staggering. Recent data from the United Nations Office of Drugs and Crime indicate that 540 different NPS have been identified worldwide as of 2014, and of course this number is expected to rise.[6]

Stimulant-like NPS Interact with Monoamine Transporters

The first stimulant-like NPS to appear in the US were so-called "bath salts" products, which flooded the recreational drug marketplace during late 2010. By early 2011, there was a dramatic rise in reports of bath salts intoxications to poison control centers and an influx of patients admitted to emergency departments with toxic exposures.[7,8] Bath salts products consist of powders or crystals that are meant to be administered intra-nasally or

Bath salts typically take the form of a white or brown crystalline powder and are sold in small plastic or foil packages labeled "not for human consumption." Sometimes also marketed as "plant food"—or, more recently, as "jewelry cleaner" or "phone screen cleaner"—they are sold online and in drug paraphernalia stores under a variety of brand names, such as "Ivory Wave," "Bloom," "Cloud Nine," "Lunar Wave," "Vanilla Sky," "White Lightning," and "Scarface."

orally to produce their psychoactive effects. While low doses of bath salts induce typical stimulant effects such as increased energy and mood elevation, high doses or binge use can cause severe symptoms including hallucinations, psychosis, increased heart rate, high blood pressure, and hyperthermia, often accompanied by combative or violent behaviors.[2,8] Deaths from bath salts overdose have been reported.[2,8]

Forensic analysis of bath salts products in 2010 and 2011 revealed the presence of three main synthetic cathinones: methylenedioxypyrovalerone (MDPV), mephedrone, and methylone.[8] These compounds are structurally related to the parent compound cathinone, an amphetamine-like stimulant found in the khat plant, *Catha edulis*. Legislation passed in 2013 placed these three synthetic cathiones into permanent Schedule I control, making the drugs illegal in the US.[9,10] Figure 1 depicts the chemical structures of bath salts cathinones compared to amphetamine. Reports of bath salts exposures

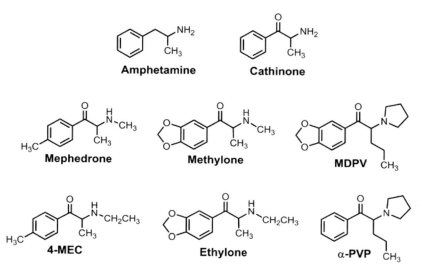

Figure 1. Chemical structures of amphetamine, cathinone, and the bath salts—athinones-mephedrone, methylone, and MDPV 4-MEC, ethylone, and lpha-PVP are replacement analogs that appeared in the recreational drug marketplace after emergency scheduling legislation was enacted in 2011 to ban mephedrone, methylone, and MDPV. (Source: Michael Baumann)

have subsided in recent years,[7] but MDPV, methylone, and other various structurally related cathinone analogs shown in Figure 1 are still present in the recreational drug marketplace.[11]

Like other stimulant drugs of abuse (e.g., cocaine and amphetamine), synthetic cathinones exert their effects by binding to "transporter" proteins on the surface of nerve cells that synthesize the monoamine neurotransmitters dopamine, norepinephrine, and serotonin. These neurotransmitters are released from nerve cells and mediate cell-to-cell chemical signaling.[12] The normal role of the transporters is to pull excess amounts of the released monoamine neurotransmitters from the "extracellular" spaces around cells and move them back into the cells that made them (a process called "reuptake"). Thus, monoamine transporters are critical for reducing extracellular concentrations of monoamine neurotransmitters. Drugs that interact with transporters can be divided into two types: 1) cocaine-like inhibitors and 2) amphetamine-like substrates. Inhibitor drugs block neurotransmitter uptake by clogging the transporter channel—much as a sock would clog a vacuum cleaner. Substrate drugs also block the transporter momentarily, but these drugs are small enough that they themselves are translocated through the transporter channel into the cell, where they trigger the efflux of neurotransmitter molecules (i.e., transporter-mediated release) into the extracellular space. The releasing action of transporter substrate drugs can be viewed as switching a vacuum cleaner into reverse, causing the dumping of intracellular neurotransmitters into the extracellular medium.

Regardless of molecular mechanism, both types of transporter drugs dramatically increase extracellular concentrations of monoamines, amplifying cell-to-cell chemical signaling in various brain circuits. Elevations in extracellular dopamine are implicated in the pleasurable and addictive properties of stimulants, whereas elevations in norepinephrine are thought to mediate cardiovascular and autonomic effects.

Pharmacologists have examined the biological effects of bath salts cathinones using a variety of methods in laboratory animals. In brain tissue and cultured cells, MDPV is a transporter inhibitor that potently blocks the uptake of dopamine and norepinephrine, with little effect on the uptake of serotonin.[13,14] Importantly, MDPV is at least ten-times more potent than cocaine in inhibiting dopamine and norepinephrine uptake. In contrast to MDPV, the transporter substrates mephedrone and methylone evoke the release of dopamine, norepinephrine, and serotonin from nerve cells.[14,15] The neurotransmitter-releasing actions of mephedrone and methylone are similar to the effects of the illicit drug MDMA. Consistent with their activities as inhibitors or substrates at dopamine transporters, administration of synthetic cathinones to rats produces dose-related increases in extracellular concentrations of dopamine in the mesolimbic system, a pathway of nerve cells implicated in pleasure and addiction.[13,15]

All synthetic cathinones investigated to date produce dose-related stimulation of locomotor activity when administered to rats or mice, probably due to enhancement of dopamine transmission. Recently, the addictive potential of synthetic cathinones has been investigated using the rat self-administration model, which is considered the gold standard for testing the abuse liability of drugs due to its high degree of predictive validity.[16]

In the self-administration studies, rats fitted with intravenous catheters are placed in chambers equipped with two levers. Presses on the "active" lever result in computer-controlled intravenous drug infusions, whereas presses on the "inactive" lever have no consequence. Under these circumstances, rats will learn to self-inject MDPV, mephedrone, and methylone, indicating that these drugs have substantial abuse potential.[16,17]

It is worth noting that MDPV) is much more potent and effective in the self-administration assay when compared to methylone. Additionally, MDPV seems more apt to produce life-threatening adverse effects in humans.[8] Even though different cathinones may be found in bath salts products, MDPV was the main compound present in blood and urine from fatal cases of bath salts overdose during the first wave of abuse in the US. It is tempting to speculate that potent inhibition of dopamine uptake by MDPV is responsible for neurological symptoms and hyperthermia in bath salts overdose cases, while inhibition of norepinephrine uptake could underlie elevations in blood pressure and heart rate.

Marijuana-like NPS Interact with Cannabinoid Receptors

Marijuana-like NPS, also known as synthetic cannabinoids, appeared in the US recreational drug marketplace in the early 2000s, and by the end of the decade were being widely used.[18,19] The initial products containing synthetic cannabinoids were marketed as "spice," "K2," or "herbal incense" and consisted of plant material laced with psychoactive compounds. Like marijuana, synthetic cannabinoids are usually smoked to produce their psychoactive effects. Low doses of synthetic cannabinoids produce marijuana-like effects, including perceptual distortions and mood elevation, but higher doses or binge use can produce serious adverse effects, including increased heart rate, uncontrolled vomiting, acute kidney injury, panic attacks, hallucinations, psychosis, and seizures.[2,3,19] Deaths from synthetic cannabinoid overdose are rare but have occurred.[2,3]

Forensic evaluation of spice and K2 products in 2010 and 2011 identified the active ingredients as the naphthoylindole JWH-018 and its analogs.[11] Figure 2 shows that JWH-018 and other synthetics are structurally distinct from naturally occurring tetrahydrocannabinol (THC),[9] the active ingredient in marijuana. JWH-018 and related synthetic cannabinoids were originally synthesized by research scientists as tools to study the function of cannabinoid-1 (CB1) and cannabinoid-2 (CB2) receptors, the cell surface receptors where THC exerts it effects.[20] Clandestine chemists hijacked the recipes for the manufacture of these synthetic drugs and made them available for misuse by humans.

Due to the public health risks posed by synthetic cannabinoids, legislation passed in 2013 placed JWH-018 and several of its analogs into permanent Schedule 1 control, making these substances illegal in the US.[9] In response to drug scheduling, many replacement analogs have appeared in the recreational drug marketplace, including cyclopropyl ketone indoles such as UR-144 and XLR-11 during 2013 and 2014, and indazoles such as AB-PINACA and AB-FUBINACA in more recent months (see Figure 2).[3,11,20] In contrast to the diminishing reports of bath salts exposures since 2011, poison control data show a sharp increase in intoxications with synthetic cannabinoids in 2015.[18]

As noted above, JWH-018 and other synthetic cannabinoids bind to and stimulate CB1 and CB2 receptors located on the ends of axons (nerve terminals) that transmit neurochemical messages from one cell to another, but the synthetic drugs are more potent and exert stronger effects than THC.[21,22] The high potency of synthetic cannabinoids at CB1 and CB2 receptors may underlie the increased propensity for these drugs to produce adverse effects when compared to marijuana.[2,3,19] Early research demonstrated that CB1 receptors are found in the brain, whereas CB2 receptors are found in peripheral organs such as the spleen.[23] More recent evidence shows that CB1 and CB2 receptors are both present in the brain, but CB1 receptors are found in much higher amounts in nervous tissue and probably mediate the psychoactive effects of THC and synthetic cannabinoids.

Under normal circumstances, CB1 receptors are stimulated by naturally occurring cannabinoid compounds (i.e., endocannabinoids) in the brain, such as anandamide and 2-arachidonoyl glycerol.[23] CB1 receptors are located on nerve terminals throughout the brain, and activation of these receptors can inhibit the release of excitatory neurotransmitters (e.g., glutamate) or inhibitory neurotransmitters (e.g., GABA).[23,24] Thus, CB1 recep-

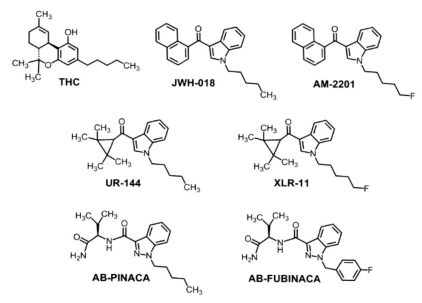

Figure 2. Chemical structures of THC, JWH-018, and common synthetic cannabinoids. AM-2201 is an analog of JWH-018 that appeared in the recreational drug marketplace after emergency scheduling legislation was enacted in 2011 to ban JWH-018. UR-144, XLR-11, AB-PINACA and AB-FUBINACA represent subsequent generations of replacement cannabinoid drugs that have appeared in the recreational drug marketplace over the past few years. (Source: Michael Baumann)

tors are modulators of the cell-to-cell signaling of other neurotransmitter systems. Endocannabinoids and cannabinoid receptors are implicated in the control of mood, appetite, pain sensation, learning, and memory. As such, drugs that interact with cannabinoid receptors would be predicted to have profound effects on brain function. The physiological role of CB2 receptors in the brain is not well understood, but the activation of CB2 receptors by cannabinoids likely contributes to the overall profile of drug effects. Additionally, it seems feasible that synthetic cannabinoids may have actions that are mediated by noncannabinoid mechanisms.

Pharmacologists have examined the effects of synthetic cannabinoids in laboratory animals using a variety of methods. The administration of THC to mice is known to produce four characteristic responses: 1) decreased motor activity, 2) reduced body temperature, 3) dulled pain sensation, and 4) a lifeless immobility, known as catalepsy.[20] These four responses, collectively known as the "tetrad," represent the hallmark profile of effects exerted by cannabinoid-type drugs. The effects of THC in the tetrad assay are dose-dependent and reversed by the CB1 receptor-blocking drug rimonabant. Not surprisingly, JWH-018 and other synthetic cannabinoids produce the tetrad of effects in mice, but are much more potent than THC.[21,22] Furthermore, synthetic cannabinoids can produce very long-lasting effects in the tetrad assay. Metabolism studies have shown that JWH-018 and its analogs are transformed by liver enzymes into several different metabolites.[25] Some of these metabolites penetrate into the brain, are potent stimulators of cannabinoid receptors, and have long half-lives. Such findings suggest that the effects of synthetic cannabinoids may be longer lasting than THC's due to their more persistent bioactive metabolites.

Unlike the situation with stimulant drugs, THC and synthetic cannabinoids are not readily self-administered by rats or mice.[20,26] Consequently, a different behavioral model, known as drug discrimination, is used to assess marijuana-like psychoactive effects in animals.[26] In the typical drug discrimination procedure, rats are trained to associate the internal cues produced by THC administration with food rewards. Importantly, drugs are administered by the scientific investigator in this paradigm and not self-injected by the rats. After repeated training sessions, rats learn to "discriminate" the internal cues produced by cannabinoid-type drugs from those produced by vehicle control treatments or other types of psychoactive substances. Studies have shown that JWH-018 and its analogs are recognized as cannabinoid-like in rats trained to discriminate THC from its vehicle,[21,26] and similar results have been found with newer synthetics, including UR-144 and XLR-11.[22] In studies where more than one synthetic cannabinoid has been evaluated, the rank order of drug potency falls in line with the potency of drugs to activate the CB1 receptor, suggesting the CB1 receptor is responsible for the observed effects.[20-23]

The Spread of NPS Presents Major Challenges

The growing popularity of NPS presents major challenges for governments, law enforcement, and public health. Controlling the influx of NPS from overseas laboratories is a complex political and economic issue that will require international cooperation among all stakeholders, especially those countries where synthetic drugs are being manufactured.[27] Even with reduced production of NPS, monitoring Internet commerce will remain problematic. In the US, drug scheduling legislation has been a primary response to the spread of NPS,[9,10] but this approach is often ineffective. For example, after the 2011 emergency scheduling action to ban MDPV and methylone in the US, law enforcement encounters involving methylone increased more than fivefold, and this substance is still present today.[11]

As mentioned previously, drug scheduling drives the emergence of new chemically distinct replacement analogs as clandestine laboratories scramble to stay one step ahead of legislative control.[1,4,11] One new analog of MDPV, known as alpha-PVP or "flakka," is a popular NPS that has wreaked havoc in Florida and other states, causing multiple deaths due to toxic overdose.[28] Analogs of methylone, such as ethylone, are now found in tablets sold in the recreational drug marketplace as "molly" (i.e., counterfeit MDMA).[29] Finally, drug scheduling hinders the ability of scientists to obtain NPS for study, thereby impeding the very research that is needed to elucidate the effects of these substances as they emerge.[30]

The abuse of NPS has placed a significant burden on health care professionals, especially those providing emergency medical care. Many cases of toxic overdose from bath salts and synthetic cannabinoids are first reported to poison control centers, and subsequently treated in hospital emergency departments.[7,18,31] As noted above, symptoms of overdose from NPS include cardiovascular effects such as increased heart rate and blood pressure, and neurological effects such as panic attacks, psychosis, and hallucinations.[2,8,19] These symptoms are often accompanied by extreme elevations in body temperature, along with combative or otherwise violent behaviors; thus, subduing and treating such patients can be a harrowing experience for hospital staff. Because the precise substance ingested by a particular patient is usually unknown, targeted treatments or receptor antagonists cannot be administered. Medical treatment is mostly supportive, with benzodiazepines to reduce cardiovascular stimulation and agitation, and whole body cooling to address hyperthermia.[31,32]

At present, most NPS are not detected by routine toxicology screens. So, analytical confirmation of synthetic drug exposure in intoxicated patients is often impossible. The fact that NPS can be used without detection is a primary driving force for the abuse of these substances in individuals subjected to routine drug testing, such as military personnel.[33] Sophisticated analytical methods for the detection of NPS and their metabolites are being developed, but such methods are not readily available in most clinical settings.[34,35] Unfortunately, forensic toxicologists are faced with the prospect of continually developing new analytical methods to keep pace with the appearance of new replacement NPS.

Widespread abuse of NPS is a complex problem with no simple solutions, and novel drugs continue to emerge at a rapid pace. NPS pose obvious health risks because there is no quality control in their production, their pharmacological effects are poorly understood, and clinical data are limited to cases from emergency room admissions. In general, the pharmacological effects of NPS seem to resemble those of the illicit drugs that they are intended to mimic; but NPS are often much more potent, and "off-target" mechanisms of action have not been established. More basic research in animal models is needed to evaluate the consequences of acute and chronic exposure to various types of NPS. Newly developed analytical methods for detecting NPS must be made widely available to assist in identifying novel substances as they emerge in the recreational drug marketplace. In the interest of public health and safety, better coordination among emergency medical personnel, forensic toxicologists, scientific researchers, law enforcement, and policymakers is essential to foster more effective responses in dealing with this evolving drug-abuse phenomenon.

2

Lithium to the Rescue

By Richard S. Jope, Ph.D., and Charles B. Nemeroff, M.D., Ph.D.

Richard S. Jope, Ph.D., is a professor in the Department of Psychiatry and Behavioral Sciences and in the Department of Biochemistry and Molecular Biology at the University of Miami School of Medicine. He obtained a Ph.D. in biological chemistry at UCLA in 1975, followed by postdoctoral training in pharmacology at UCLA. During his postdoctoral fellowship, he began to study the mechanism of action of lithium, publishing his first research paper on lithium in 1978 in the *New England Journal of Medicine*, which has been followed by over 100 papers and book chapters focused on lithium's actions. Upon moving to a faculty position at the University of Alabama at Birmingham School of Medicine, Jope obtained funding from the NIH to study lithium in 1984, a project that continues to be funded to identify the therapeutic actions of lithium.

Charles B. Nemeroff, M.D., Ph.D., is the Leonard M. Miller Professor and chairman of the Department of Psychiatry and Behavioral Sciences, and director of the Center on Aging, at the University of Miami School of Medicine. His research has focused on the pathophysiology of mood and anxiety disorders and the role of mood disorders as a risk factor for major medical disorders. Nemeroff received his M.D. and Ph.D. degrees in neurobiology from the University of North Carolina (UNC) School of Medicine. After psychiatry residency training at UNC and Duke University, he held faculty positions at Duke and at Emory University before relocating to the University of Miami in 2009. He has served as president of the American College of Psychiatrists (ACP) and the American College of Neuropsychopharmacology. He has received the Kempf Fund Award for research development in psychobiological psychiatry; the Samuel Hibbs Award, Research Mentorship Award, Judd Marmor Award, and Vestermark Psychiatry Educator Award from the American Psychiatric Association (APA); and the Mood Disorders Award, Bowis Award, and Dean Award from the ACP. He is the co-editor-in-chief of the *Textbook of Psychopharmacology*, published by the APA. He is a member of the National Academy of Medicine.

 *Lithium, an element that Mother Nature has put in some drinking water sources, has been used for its curative powers for centuries. Today, it's given in capsule form as a mood stabilizer for bipolar disorder and depression. New research, however, reveals its role as a neuroprotector, and suggests that a better understanding of the role enzymes modulated by lithium play could lead to new treatments for Alzheimer's disease, Parkinson's disease, multiple sclerosis, and other neurodegenerative disorders.*

IN THE POPULAR TELEVISION SERIES *Homeland*, lithium is among the medications used by the bipolar disorder-afflicted main character, Carrie, played by Claire Danes. When Carrie is off her meds, her judgment is compromised, and chaotic situations tend to worsen. At one point last season, Carrie went off her lithium because she mistakenly believed it would help her improve her concentration.

Although most people are better acquainted with lithium through shows such as *Homeland* and other forms of popular culture, others know of it because of its use in a clinical setting. Individuals typically take lithium for bipolar disorder, a condition that involves cycles of mania and depression and is often referred to by its previous name—manic-depressive illness.

TV stereotypes aside, growing evidence suggests that low levels of lithium may strengthen the brain's resilience to stress and disease. There is now a substantial understanding of how lithium can strengthen and protect the brain.

Few people realize that lithium is the simplest molecule in any pharmacopoeia. Lithium is simply a small, positively charged element, similar to sodium. In nature (found in groundwater, for example) and in medicine, charged molecules are neutralized with a counterbalancing molecule. As a result, lithium is attached to a negatively charged molecule, such as chloride, to form the uncharged compound lithium chloride.

A Long History

Lithium was used inadvertently in medicine long before it was identified as a unique element in the early 1800s. It was used for maladies of many kinds—from centuries of "taking the waters" at Marienbad, Vichy, or Baden Baden, and other sources of "healthy waters," which often were naturally enriched in lithium, to the more recent popularity of ingesting "lithium-fortified" beverages for improving health and immunity in the 20th century. The popular soda 7-Up was supplemented with lithium until 1950, for example.

In the 1950s, when clinical trials of lithium were first conducted, it became clear that lithium had the capacity to stabilize mood in about half of the patients with bipolar disorder. Indeed, lithium remains the "gold standard" of treatment for bipolar disorder to which all subsequent treatments are compared. It is generally considered the most effective mood stabilizer for the severe disorder's manic and depressive components. Lithium's clinical efficacy set in motion the current research effort to better understand lithium's mechanisms of action. Its untapped potential may well be considerable and beneficial in the treatment of several other diseases.

Brain Protector

Understanding the causes of diseases and how therapeutic drugs work is a long, arduous process. Research into how lithium stabilizes mood in bipolar disorder patients has been no exception. Many different effects of lithium were found by researchers studying different kinds of cell types, organs, and organisms. However, the difficulty lay in determining which of these actions was important for its therapeutic effect.

In 2002, Drs. Husseini Manji and Carlos Zarate[1] organized a wonderful little conference to which they invited leading mood disorder researchers to present their best ideas about how mood stabilizers such as lithium work. Not surprisingly, a large number of different possibilities were presented. Our own contribution was to present evidence that an important action of lithium was its neuroprotective effect. Recognized for several years, this action was mediated by its inhibition of an enzyme called glycogen synthase kinase-3 (GSK3).[2] The idea was based on the discovery by others that lithium inhibits the activity of GSK3,[3] and the discoveries that GSK3 inhibition protects neurons from a wide variety of insults, including oxidative stress, impaired mitochondrial functions, DNA damage, and excitotoxicity.[4,5] GSK3 is an enzyme in a class known as a kinases that transfer phosphate groups to other proteins to regulate their activities.

Not surprisingly, most of our colleagues were skeptical and followed other directions in their research. This skepticism was partially based on the fact that the neuroprotection offered by lithium and other inhibitors of GSK3 was first detected in experiments measuring the death of neurons, whereas psychiatrists widely believe that neurons in patients with bipolar disorder do not die. Unappreciated by many at the time was the concept that neurons undergo a number of degenerative changes that impair their functions long before they actually die. Examples of impairments that often occur well before cell death are weakening or loss of connections (synapses) between neurons that are required for communication, pruning of cell processes such as dendrites and nerve terminals, changes in the function of the mitochondria that produce the energy used by cells, stress responses in cells (such as in the endoplasmic reticulum stress), loss of the protective myelin sheaths around neuronal axons, and DNA damage.

Over time, it became evident—and has become accepted in the past decade—that although lithium can sometimes stave off neuronal death, probably most of its clinical benefit comes from its prevention of these impairments that precede neuronal death. It is also important to note that lithium is not only neuroprotective but is also one of the psychopharmacological agents that increases new neuron production (neurogenesis) in the hippocampus, a critical brain area for learning, memory, and stress responses.[5]

Many people wonder how a small element such as lithium can be neuroprotective and interact with only a very small number of proteins such as GSK3. Recent studies of the structures of proteins reveal that a few contain a small pocket in which lithium fits perfectly.[6] When lithium is in the pocket, it acts like a stone in a cog, blocking the normal workings of the protein. GSK3 is such a protein. When lithium is in the pocket of GSK3, it blocks the actions of GSK3 and consequently alters the actions of numerous proteins that are regulated by GSK3.

One of the effects of GSK3's phosphorylation activities is to facilitate signals that cause neurons to die.[4] This is a normal process during development when the brain needs to rid itself of excessive neurons. Unfortunately, in many neurodegenerative conditions, including Alzheimer's disease, Parkinson's disease, traumatic brain injury, and ischemia following stroke, the mechanisms causing neuronal death are abnormally activated. The large number of people afflicted with these conditions makes it critical to better understand lithium's neuroprotective effects to determine if it may be beneficial in such cases.

Biochemical Mechanisms

That lithium can produce many effects in cells was somewhat surprising for a while, given the small number of proteins with which lithium directly interacts. However, lithium's unusual breadth of activity appears to result from its inhibition of GSK3, which itself regulates the functions of more target proteins than any other kinase through the phosphorylation mechanism noted above.[7] In fact, GSK3 is known to phosphorylate, and thereby regulate, more than 100 proteins. A good number of these play roles in regulating neuronal resilience to stress.

GSK3 is activated when neurons are stressed by any of a large variety of mechanisms, such as oxidative stress, endoplasmic reticulum stress, DNA damage, and exposure to toxic chemicals or proteins.[7] Activated GSK3 then has an increased influence on cellular proteins and functions, often promoting the deleterious effects of stress on neuronal functions that in severe conditions can cause neuronal death. Lithium can bolster the resilience of neurons by inhibiting GSK3 to put a brake on the deleterious effects of stress and toxic substances.

One large class of proteins regulated by GSK3, and hence by lithium, is transcription factors.[7] Transcription factors regulate the expression (i.e., transcription) of genes, thereby regulating the levels of proteins that are present in cells. GSK3 is now known to regulate the actions of more than 25 different transcription factors, thereby exerting a tremendously large effect on regulating the levels of proteins in neurons. For example, GSK3 inhibits the transcription factor called CREB—which otherwise contributes to cellular resilience and learning and memory, in part by inhibiting the expression of the signaling protein BDNF. Lithium effectively frees CREB to do its beneficial work. Lithium also increases resistance to oxidative stress by reversing GSK3's inhibition of the transcription factor Nrf2. Moreover, lithium counters GSK3's activation of the transcription factor p53, which promotes cell death in response to certain stresses. Thus, lithium's inhibition of GSK3 can bolster the actions of CREB and Nrf2, and diminish the action of p53, to promote neuronal resilience.

Lithium's inhibition of GSK3 also contributes to an array of other neuroprotective actions too numerous to discuss. Well-documented outcomes include axonal regeneration, improved mitochondrial function, remyelination, and the generation of new neurons in adult mammalian hippocampus (neurogenesis). These neuroprotectant and neurorestorative actions of lithium each depend on modifications of different proteins in each process, often mediated by GSK3. Thus, lithium has the capacity to provide neuroprotection in many conditions that exert distinct deleterious effects on neurons.

Lithium's Vast Potential

While lithium is well-established as an effective therapy for many patients with bipolar disorder, recent research in rodent models has found evidence that it also may provide clinically significant neuroprotection in other conditions.

As mentioned above, a number of toxic proteins and chemicals that adversely affect, or stress, neurons are capable of activating GSK3 in neurons, and several of these have been linked to diseases. For example, Alzheimer's disease appears to be caused in part by abnormal accumulation of a protein called Aβ, which is the primary component of amyloid plaques that develop in the brains of patients with Alzheimer's disease. Aβ has been shown to activate GSK3, which in turn causes abnormal phosphorylation of a protein called tau, the primary component of neurofibrillary tangles in Alzheimer's disease brains.[8] Thus, GSK3 is intimately involved in the neuropathological hallmarks of Alzheimer's disease, as well as in the death of neurons.

An exciting possibility is that lithium may delay the devastating pathology and neuronal loss that occurs in Alzheimer's disease. This possibility is supported by evidence that dementia is less prevalent in patients with bipolar disorder who have been taking lithium for long periods of time, though this is only a correlative association.[9] The strong theoretical basis for lithium therapy led to some early trials of lithium in patients with Alzheimer's disease, which revealed mild beneficial effects of lithium.[10,11] However, the effects of lithium may be improved by initiating treatments very early in the disease, before neurons are damaged beyond repair. In determining therapeutic effectiveness, it's also likely that lithium and other treatments need to be given for several years to test the beneficial effects in a clinical trial—although a long trial is a very difficult task.

The apparently abnormal activation of GSK3 in neurons is seen also in Parkinson's disease,[1] stroke, and traumatic brain injury. Using lithium to inhibit GSK3 may therefore have some benefit in these disorders as indicated by its protective effects in animal models of these conditions.[12-14]

Another important neuroprotective mechanism of lithium's inhibition of GSK3 appears to be the control of inflammation.[15] Inflammation is associated with systemic disorders, such as rheumatoid arthritis and Crohn's disease, and is also evident in many brain diseases, including Alzheimer's disease, Parkinson's disease, and stroke. It is also particularly important in multiple sclerosis.

Injury and infections are known to activate inflammation, but less well-known is the fact that stress can activate inflammation. Inflammation is a double-edged sword. In some conditions it is essential for surviving pathogenic infections. However, uncontrolled or aberrant inflammation, which may occur in response to stress, can be detrimental to neuronal functions. Lithium is a surprisingly effective controller of inflammation, at least in part by its inhibition of GSK3. In animal models and studies in cells, lithium and other GSK3 inhibitors strongly reduce inflammation.[16] This anti-inflammatory effect of inhibiting GSK3 was sufficient to induce remission of the clinical symptoms in animal models of multiple sclerosis.[17] Thus, the anti-inflammatory actions of lithium likely make an important contribution to its neuroprotective capacity.

Lithium may also be effective in treating Fragile X syndrome (FXS), for which no effective therapy exists. FXS is caused by inherited gene mutations that result in intellectual disability and other behavioral disturbances. GSK3 is abnormally active in the brain of animal models of FXS.[18] Inhibition of GSK3 with lithium or other drugs effectively reverses many of the abnormal characteristics in the mouse model of FXS, including reversing impairments in some forms of learning and memory. Intriguingly, lithium is the only treatment as yet found to improve cognition in patients with FXS.[19] The neuroprotective mechanisms of lithium in FXS remain to be identified but are thought to involve improving synaptic communication between neurons.

Finally, a particularly intriguing finding is that lithium in therapeutic doses in bipolar patients significantly reduces suicide rates. Even small amounts of lithium may effectively diminish suicidal behavior.[20,21] Several epidemiological studies found that locales with higher natural levels of lithium in the drinking water have lower suicide rates than those with less lithium.[22] Although only correlative, these data are consistent with studies in patients that demonstrate that therapeutic levels of lithium reduce suicidal behavior in bipolar patients. How lithium alters neurons to diminish suicide is currently under intense investigation, as there are few other interventions available that reduce suicide, most notably the atypical antipsychotic clozapine.

We should also note that there is also currently much interest in the ability of very low doses of lithium to work synergistically with other therapeutic drugs. For example, Chuang and colleagues pioneered the notion that low doses of lithium synergistically cooperate with low doses of another mood stabilizer, valproate, or the investigational antidepressant ketamine.[23] Investigators have reported particularly strong damage-preventing effects of combined low doses of lithium and valproate in mouse brains after experimental stroke.

Possible Limitations

New is fashionable in clothes, cars, and even drugs—people frequently crave the latest thing. Yet when the need arises, we often rely on the wisdom of our ancestors for guidance. Folk medicine has touted the benefits of lithium for many years, and scientific research is now providing answers to how lithium protects the brain from a wide range of conditions, and for which diseases lithium therapy may be most effective.

While our story has focused on neuroprotective effects of lithium that derive from its inhibition of GSK3, this focus is not meant to overlook the possibility that other targets of lithium may also contribute to its neuroprotection. It is also important to note that lithium treatment is not without limitations, as side effects can limit its utility; doses must be carefully monitored because high therapeutic levels can be toxic.

It is also important to emphasize that, except for bipolar disorder, lithium is likely to be most effective when used in conjunction with other therapies that are aimed at targets specific to each disease, such as the mechanisms that generate toxic proteins and chemicals. Although lithium provides neuroprotection, the effects of many insults are too powerful to be overcome by any individual known treatment. As a result, therapies aimed at reducing the toxic insults combined with lithium's neuroprotective abilities may prove to be most useful to ameliorate conditions involving loss of neuronal functions.

3

The Malignant Protein Puzzle

By Lary C. Walker, Ph.D., and Mathias Jucker, Ph.D.

Lary C. Walker, Ph.D., is associate professor of neurology and research professor of neuropharmacology and neurologic diseases at the Yerkes National Primate Research Center, and associate director of the Alzheimer's Disease Research Center at Emory University. Walker earned his Ph.D. at Tulane University, and is an authority on the pathogenesis of Alzheimer's disease and the role of abnormal proteins in neurodegeneration. With Mathias Jucker, Ph.D., he has pioneered studies of how disease-related proteins are induced to misfold, aggregate, and spread in the brain. Walker is currently working to understand the similarities and differences between Alzheimer's disease and other neurodegenerative disorders, and why humans as a species are especially vulnerable to Alzheimer's disease.

Mathias Jucker, Ph.D., is a professor at the Hertie Institute for Clinical Brain Research at the University of Tübingen and the German Center for Neurodegenerative Diseases in Tübingen. He is head of the university's Department of Cellular Biology of Neurological Diseases and spokesperson of the Graduate School of Cellular and Molecular Neuroscience in Tübingen. Jucker studied neurobiology at the Eidgenössische Technische Hochschule in Zurich and completed his Ph.D. there in 1988 before working as a postdoc and research scientist at the National Institute on Aging in Baltimore. He relocated to the University of Basel as a junior professor (START fellow) and, in 2003, moved to the Hertie Institute.

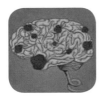 *When most people hear the words "malignant" and "brain,"-cancer immediately comes to mind. But our authors argue that proteins can be malignant too, and can spread harmfully through the brain in neurodegenerative diseases that include Alzheimer's, Parkinson's, CTE, and ALS. Studying how proteins such as PrP, amyloid beta, tau, and others aggregate and spread, and kill brain cells, represents a crucial new frontier in neuroscience.*

ON THE AUSTRALIAN ISLAND OF TASMANIA around 20 years ago, a disfiguring, fatal cancer of the face was reported to be rapidly spreading among Tasmanian devils. The disease, known as devil facial-tumor disease, happens to be an extraordinary instance of infectious cancer. It is caused not by a virus but by the direct transfer of cancer cells from one devil to another, possibly through biting.[1] And it is not unique to devils; other examples of unusual infectious cancers have been described in species such as dogs[2] and clams.[3]

These curious cases reveal that some cancer cells can "infect" receptive hosts, but they by no means indicate that all malignancies should be treated as infectious diseases. The great majority of cancers arise within the body of the host when normal cells transform and proliferate uncontrollably. Infectious cancers do, however, highlight the impartial resourcefulness of biology in both health and disease.

We now believe that many of the neurodegenerative diseases that increasingly plague modern humans bear an intriguing similarity to cancer, except that the disease agents that proliferate in these brain disorders are not transformed cells, but rather transformed proteins that have folded into the wrong shape. Such "malignant" proteins are key players in such devastating diseases as Alzheimer's, Parkinson's, Huntington's, frontotemporal dementia, chronic traumatic encephalopathy (CTE), amyotrophic lateral sclerosis (ALS), and Creutzfeldt-Jakob disease (CJD). Most of these maladies are thought to be noncontagious under ordinary circumstances, but CJD and its variants have been transmitted to humans by tainted meat, cannibalism, and tissue transplants, and research suggests that other disease-linked malignant proteins can in some circumstances transmit their properties from one organism to another.

As in the case of cancerous cells, though, these rogue proteins almost always emerge and propagate within the affected host. Once the misfolded proteins gain a foothold in the nervous system, they effectively compel normal versions of the same protein to adopt the same malformed state. In this faulty configuration, the proteins stick to one another and, in a molecular chain reaction, structurally corrupt like molecules that are generated in the course of normal cellular metabolism. In many instances, the final products of this process are clumps of the protein called amyloid (Figure 1). Central to this process is what we call seeded protein aggregation (seeding for short), a surprisingly common disease mechanism that first came to light with the discovery that protein seeds called prions can act as infectious agents.

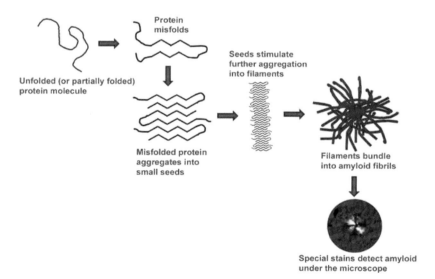

Figure 1. The steps leading from a single protein molecule to clumps of amyloid. The misfolded proteins act as seeds that accelerate the crystallization-like disease process. The seeds can vary in size, and very small assemblies called oligomers can be particularly toxic to cells. In prion diseases, the seeds can sometimes be transferred from one organism to another ("infection"). (Image courtesy of Lary Walker/Mathias Jucker)

The Prevailing Prion

The improbable tale of infectious proteins began in the 1730s, when reports of a slowly progressing and ultimately fatal disease of sheep first appeared in the European scientific literature. British farmers called the disease scrapie because affected sheep were seen to scrape the wool from their skin by compulsively rubbing against farmyard objects. The farmers suspected even then that scrapie was contagious, but it wasn't until the 1930s that transmission of the disease was demonstrated experimentally by Jean Cuillé and Paul-Louis Chelle in France.[4,5] To establish infectivity, Cuillé and Chelle injected homogenized nervous tissue from scrapie-afflicted donor sheep into healthy host sheep. Infectious illnesses usually emerge within days or weeks, but earlier experiments had failed to demonstrate transmission of scrapie in this timeframe. Cuillé and Chelle, however, were patient; the sheep that they injected with scrapie-tainted tissue finally succumbed to the disease more than a year later.

Thus began a long and prickly debate about the nature of the scrapie agent: What kind of infectious pathogen causes disease only after months or years of incubation? Furthermore, infections usually announce their presence with inflammation and fever, yet scrapie showed no such signs. The term "slow virus" was adopted by many, but evidence gradually mounted that the culprit was not a virus at all, but rather, just possibly, an infectious protein.

Interest in the problem intensified in the 1960s when D. Carleton Gajdusek and his colleagues made the Nobel Prize-winning discovery that, like scrapie, two rare human neurodegenerative diseases, kuru and Creutzfeldt-Jakob disease, are transmissible with very long incubation periods.[6] By then it was becoming clear that the agent of infection was strange indeed. Radiobiological experiments performed by Tikvah Alper strongly suggested that the agent did not require nucleic acids to replicate,[7] and the mathematician John Griffith described, prophetically, how a protein-only agent might multiply using the host's genetic machinery to generate more protein.[8] In 1982, Stanley Prusiner crystallized the protein-only concept (and enraged its opponents[9]) by naming the scrapie agent a "proteinaceous infectious particle," or "prion." In subsequent years, Prusiner's group, along with a growing cadre of allies, amassed persuasive experimental support for the prion concept, for which Prusiner was awarded the Nobel Prize in 1997. Although echoes of the old debate about the causative agent still sometimes find their way into print,[10] the prion paradigm has prevailed, and today it is evolving into a far-reaching new concept of disease.[11-14]

Assembling into Amyloid

The prions of CJD and scrapie are submicroscopic assemblies of a natural mammalian protein known as prion protein, or PrP. Prions consist of misfolded versions of PrP that can seed the formation of similar assemblies by a process resembling the seeded crystallization seen in some chemical reactions[15] (Figure 1). In this sense, PrP prions can be viewed as malignant proteins that multiply and spread within the nervous system, eventually causing neurological dysfunction and death. In humans, PrP prions trigger progressive, fatal neurodegenerative disorders that include CJD, kuru, Gerstmann-Sträussler-Scheinker disease, and fatal insomnias.[16] In nonhuman species, the PrP prion diseases include scrapie, bovine spongiform encephalopathy ("mad cow disease"), and chronic wasting disease of North American deer, elk, and moose.[16,17]

Before the discovery of prions, these diseases often were called spongiform encephalopathies due to the appearance of sponge-like vacuoles in the brain (Figure 2). Despite being caused by just one type of protein—PrP—different prion diseases display remarkably different clinical and pathological signs, and these differences appear to be related to distinct molecular features of the prions.[18] The exact mechanisms by which nervous tissues are damaged in prion diseases (and other neurodegenerative disorders) remain incompletely understood—and that poses a pressing challenge for future research.

The enigma of infectious proteins was deepened with the discovery that PrP prion diseases also can be caused by mutations in the gene that encodes PrP.[9,12] These hereditary forms of prion disease, ironically, helped to establish the protein-only hypothesis of the infectious agent, as they con-

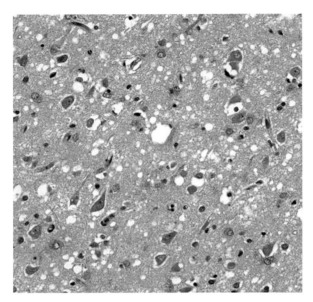

Figure 2. Spongiform vacuoles (here seen in the microscope as holes) in a thin slice of the brain of a patient who died of Creutzfeldt-Jakob disease, a prion disease. Neurons and the nuclei of other brain cells are darkly stained. (Image courtesy of Lary Walker/Mathias Jucker)

firmed the importance of normally generated PrP in the creation of new prions. In other words, hereditary (and presumably also spontaneous) PrP prion disease begins when prions are formed by the misfolding of PrP that is generated by cells inside the body. Infectious prion disease, in contrast, results when external prions invade the body of the host. But in all cases, once the process begins, the abnormal prion protein accumulates in the nervous system and triggers the impairment and death of neurons. As abnormal PrP aggregates build up in the brain, they sometimes clump into distinctive masses of amyloid.[18]

This attribute of misfolded PrP—its ability to assemble into amyloid—furnishes telling clues to the nature of the prion. Surprisingly, amyloid-forming proteins also characterize more common age-related neurodegenerative diseases, such as Alzheimer's disease. While amyloid is obvious under the light microscope (see Figure 1), the amyloid itself may only be the tip of the iceberg; the aberrant proteins often form assemblies that are not amyloid in the strict sense of the word, and in many instances, much smaller clusters of misfolded protein molecules called oligomers have been found to be quite toxic to cells.

Other Plaques and Tangles

Intriguing similarities, such as the presence of amyloid and relentless decline of brain function, suggested to Gajdusek, Prusiner, and others that the prion diseases might yield insights into the cause of Alzheimer's, Parkinson's, and other human neurodegenerative disorders. Virtually all of these maladies involve the appearance of characteristic protein deposits in the brain. For example, in Alzheimer's disease (the most frequent cause of dementia), a protein called amyloid beta ($A\beta$) aggregates to create the "senile plaque" formations seen in the gray matter of all Alzheimer's brains (Figure 3), as well as cerebral amyloid angiopathy, a buildup of amyloid in the walls of brain blood vessels. Another protein called tau also can adopt the amyloid structure, forming neurofibrillary tangles in Alzheimer's (Figure 3). While both plaques and tangles are necessary to drive Alzheimer's disease, the prime mover in the degenerative cascade appears to be $A\beta$.[19] In Parkinson's disease, yet another protein known as α-synuclein assembles into intracellular amyloid clumps called Lewy bodies. The list of diseases and their misshapen proteins continues to grow.[11] In each disease, the flawed proteins are associated with distinctive signs and symptoms. But are they, like PrP prion disease, transmissible?

In the 1960s, Gajdusek's group began a massive study to address this very question. Specifically, they wanted to know if non–PrP neurodegenerative diseases such as Alzheimer's are transmissible to nonhuman primates. The outcome was essentially negative.[20] In Great Britain, however, a team led by Rosalind Ridley and Harry Baker reported in the early 1990s that $A\beta$ plaques and cerebral amyloid angiopathy are increased in the brains of marmosets several years after injection of Alzheimer brain homogenates into the brain.[21] The actual agent that precipitated these amyloid deposits, however, remained uncertain.

These researchers logically used nonhuman primates to assess the potential transmissibility of Alzheimer's disease, since close evolutionary relatives are most likely to manifest the same type of disease. Such experiments, however, were hampered by issues of time and cost. Normal laboratory mice and rats were not suitable for these experiments because the chain of amino acids that makes up rodent Aβ differs from that in humans and monkeys; for that and perhaps other reasons, rats and mice do not naturally develop amyloid deposits in the brain as they grow old. In the mid-1990s, however, genetically engineered mouse models were introduced that make human-sequence Aβ. These "transgenic" mice generate amyloid plaques within a matter of months, and thus were widely adopted as the first practical animal models for studying Alzheimer-like Aβ aggregation in the brain.

Testing a Hypothesis

With this important new tool in hand, the two of us set out to test the hypothesis that Aβ-amyloid can be induced to form in the brains of transgenic mice by a mechanism similar to the infectivity of PrP prions. In our earliest studies, we homogenized brain tissue from Alzheimer patients, spun it briefly in a centrifuge to remove larger debris, and injected a small amount (usually 1 to 4 millionths of a liter, or microliters) of the clear extract into the brains of transgenic mice expressing human-sequence Aβ. After an incubation period of several months, the mice began to develop Aβ plaques and cerebral amyloid angiopathy in the injected region, similar in many ways to the Aβ amyloid pathology seen in Alzheimer's. Subsequent experiments in our labs and others have shown that the seeding agent is indeed aggregated Aβ.[11]

The mice did not develop full-blown Alzheimer's disease, which, to the best of our current knowledge, occurs only in humans. Research has shown, however, that at the molecular level, Aβ seeds resemble PrP prions in virtually every way: They consist solely of a particular protein; the seeds vary in size; they resist destruction by high temperature or formaldehyde; they can spread within the brain and to the brain from elsewhere in the body; and different seed structures have different biological properties (variants that are referred to as strains).[11,14]

More recently, numerous elegant studies have found that proteins involved in other neurodegenerative diseases also have prion-like properties. These proteins include tau (which forms neurofibrillary tangles in Alzheimer's disease, CTE, and many other disorders), α-synuclein (which forms

Lewy bodies in Parkinson's disease, Lewy body dementia, and multiple system atrophy), huntingtin (which forms inclusion bodies in Huntington's disease), and several proteins with prion-like properties that accumulate in such disorders as ALS and frontotemporal dementia.[11,12,22-25]

Are Neurodegenerative Proteopathies Infectious?

A growing awareness of the similarities between PrP prions and other protein seeds has revived speculation that Alzheimer's and other neurodegenerative diseases might be infectious. This question gained recent prominence with a report from a team led by Sebastian Brandner and John Collinge in Great Britain showing that at least one facet of Alzheimer's disease—Aβ-amyloid formation—appeared to be induced in patients who were treated as children with human growth hormone in order to correct short stature.[26] It was discovered in the mid-1980s that some of the growth hormone used for treatment, which had been isolated from large batches of human pituitary glands collected at autopsy, was contaminated with PrP prions. As a result, some recipients died of Creutzfeldt-Jakob disease many years after their growth hormone treatments had ceased. Brandner, Collinge, and co-workers were able to examine the brains of eight of them who were 36 to 51 years-old at the time of death. In addition to PrP prion pathology, four of the patients also had substantial Aβ accumulation in plaques and cerebral blood vessels, and two others had sparse Aβ deposits.

The appearance of Aβ plaques and vascular amyloid in people at such a young age is quite unusual. The findings strongly suggest that some batches of growth hormone were contaminated with Aβ seeds in pituitary glands that were inadvertently collected from Alzheimer patients. Remarkably, none of the eight subjects had evidence of neurofibrillary tangles, the other defining brain abnormality in Alzheimer's disease. Because all of them had died of prion disease, we cannot know whether they eventually would have developed Alzheimer's. If so, the incubation period would likely be at least as long as that of prion disease.

This presumed transmission of Aβ-amyloidosis to humans occurred under extraordinary circumstances—repeated, long-term injections of a hormone derived from pooled human pituitary glands. By sheer good luck, recombinant growth hormone (produced by genetically modified bacteria) became available in 1985, just at the time when the cadaver-derived hormone was confirmed to be contaminated with PrP prions. The recipients were quickly switched to this safer version of the agent. Strangely (or perhaps not), a black market continued to flourish for cadaver-derived growth hormone, sustained in large part by bodybuilders and other athletes; the cadaver-derived hormone is indistinguishable from that produced by the recipient, and thus is difficult to detect in doping tests.[27]

Fortunately, most of the patients treated with growth hormone prior to 1985 have not developed prion disease. It will be important to follow them in the coming years to determine whether they are at higher risk of Alzheimer's disease and other neurodegenerative disorders involving protein seeds. Interestingly, a team of Swiss and Austrian researchers recently reported a similar induction of Aβ deposition in CJD patients many years after they had received transplants of PrP prion-contaminated dura mater that had been harvested from human cadavers.[28] These studies by no means indicate that Alzheimer's disease can be transmitted from person to person under everyday circumstances; rather, they do provide the first evidence that the aggregation of a protein other than PrP might be induced in the human brain by exogenous seeds. Just as cancer cells can occasionally transmit disease from one animal to another, the same appears to be true—under exceptional circumstances—for some pathogenic protein seeds.

Promise and Pitfalls of the Prion Paradigm

In Alzheimer's and other non-PrP neurodegenerative diseases, malignant protein seeds arise from normally generated proteins inside the body, just as malignant cells stem from normal cells in cancer. Nature also has exploited the prion mechanism for beneficial ends; proteins that form prion-like aggregates handle functions ranging from information transfer in yeast[29] to the storage of peptides[30] and the consolidation of memory[31] in mammals. In light of these discoveries, we have argued that the term "prion" should be redefined as a "proteinaceous nucleating particle" to stress the molecular process of seeded protein aggregation (nucleation) that is common to all of these phenomena.[14] By removing the disquieting word "infectious," the new definition accommodates the many instances in which such proteins are not infectious by any customary definition of the term.

Recent research has brought into clearer focus the devastating role of malignant proteins in diverse diseases. Infectivity is undoubtedly an important characteristic of PrP prions, particularly in some nonhuman species. Chronic wasting disease, for example, is rapidly spreading among members of the deer family in western North America.[32] In humans, though, most cases of PrP prion disease do not result from infection. Because prions achieved notoriety largely due to their infectivity, this peculiar feature has colored our view of all cases of PrP prion disease, whether they are caused by infection or not. In light of the history of the prion concept, one has to wonder how we would view cancer if the first malignancy discovered had been devil facial-tumor disease, and only later did we learn that most cancerous cells actually develop within the body of the affected organism. Would we now consider all cancers to be potentially infectious? And at what cost to the patients and those who care for them? Our perception of disease, and the language we use to define it, must continually adapt to new information. By highlighting the molecular properties of malignant proteins, the evolving prion concept will help to guide future experimental strategies for defeating a multitude of intractable conditions.

4

Imaging the Neural Symphony

By Karel Svoboda, Ph.D.

Karel Svoboda, Ph.D., who was born in the Czech Republic, received a doctoral degree in biophysics from Harvard University and performed his postdoctoral work at Bell Laboratories. From 1997 to 2006, he was a professor at Cold Spring Harbor Laboratory, and he has been a Howard Hughes Medical Institute (HHMI) investigator since 2000. At Cold Spring Harbor Laboratory, the Svoboda lab developed and exploited methods to track synaptic transmission and plasticity at the level of individual connections, even in the intact brain. In 2006, the Svoboda lab moved to the newly established HHMI's Janelia Farm Research Campus, where Svoboda is a group leader. His current work focuses on how neocortical circuits produce our perception of the world and our actions within it. He also has a long-standing interest in new biophysical and molecular methods for brain research. In 2015, he was awarded the Brain Prize from the Grete Lundbeck European Brain Research Foundation. Svoboda is a member of the National Academy of Sciences (USA).

 *Since the start of the new millennium, a method called two-photon microscopy has allowed scientists to peer farther into the brain than ever before. Our author, one of the pioneers in the development of this new technology, writes that "directly observing the dynamics of neural networks in an intact brain has become one of the holy grails of brain research." His article describes the advances that led to this remarkable breakthrough—one that is helping neuroscientists better understand neural networks.*

SOUTH AFRICAN BIOLOGIST AND Nobel Prize winner Sydney Brenner once said "progress in science depends on new techniques, new discoveries, and new ideas—probably in that order." The mammalian brain is a case in point. It is an incredibly complex organ, and major new insights into its function have typically followed on the heels of novel scientific instruments and new methods. To illustrate the brain's complexity, scientists often use numbers: One cubic millimeter of gray matter, about the size of a grain of rice, contains 100,000 neurons (the overall number of neurons in the brain is similar to the number of stars in the Milky Way). Each of these neurons connects to approximately 1,000 other neurons at synapses, which are tiny communication channels between brain cells.

But numbers alone do not do justice to the brain. Neurons fall into multiple classes, a dozen or more clustered in each of the approximately 1,000 regions of the brain, and they take elaborate shapes. They sprout large tree-like appendages at both ends—dendrites, which receive inputs, and axons, which send outputs. These diverse neuron types hook up in extensive and intricate neural networks that link our senses to our musculature and produce intelligent behavior.

Similar to digital computers, these neural networks use electrical signals to process information. Neural computations can happen faster than the blink of an eye (after all, the blink reflex is caused by electrical signals zipping through a three-neuron chain), so that we can act quickly in a rapidly changing world. However, the brain also accumulates information gradually as we learn. Our memories allow our brains to construct a model of the world, which influences quick, millisecond processing. Memories can last a long time; for example, most of us remember our first day of school or our first kiss. This is because neurons process information over times of milliseconds to years.

Meanwhile, our abilities to navigate dynamic environments and store information are based on a sort of symphony of electrical signals that neurons produce, each with its own melody, timbre, and rhythm. Each neuronal "tune" organizes into orchestras, which ultimately conduct our perception of the world and our actions within it. Until the turn of the century, scientists had only one option to interrogate neurons in the intact brain—insert a wire into the brain, close to a single neuron. But the isolated electrical readout from a single neuron is a pale reflection of the grander symphony. Even using a far more elaborate setup, such as magnetic resonance imaging, one of only a few techniques routinely used in humans, only shows average activity in large regions of the brain, not nearly detailed enough to reveal neuronal signals. Directly observing the dynamics of neural networks in an intact brain has become one of the holy grails of brain research.

Now, through a confluence of optical physics, protein engineering, and molecular biology, it is possible to observe directly the neural symphony in the intact brain. A method called two-photon microscopy allows researchers an unprecedented view of the living brain in action, from individual synapses to entire neurons and neural circuits.

Imaging Neurons: Why Two Photons Are Better than One

Neurons and their synapses are microscopic. To study them, scientists often use biological microscopes that utilize a molecular process called fluorescence to generate detailed images. In fluorescence microscopy, scientists add special molecules to the tissue, either as chemicals or by reprogramming the genomes of cells to coax them into producing fluorescent proteins. Samples are then illuminated with colored light that is absorbed by the fluorescent molecules, which, in turn, emit fluorescence. The molecules absorb

Figure 1. Diagrams illustrating different modes of fluorescence. Left, standard fluorescence. A blue illumination photon (1) excites a green fluorescent molecule (A). The molecule subsequently emits a green photon (2). Right, two-photon excitation of fluorescence. Two infrared illumination photons (3) combine together to excite the green fluorescent molecule (A). The infrared photons have to arrive nearly simultaneously for two-photon excitation to occur. The molecule subsequently emits a green photon (2), identical to the standard case.

the high-energy illumination photons, or particles of light, and then in response emit lower-energy fluorescence photons (Figure 1). The fluorescent substances can then be detected with exquisitely sensitive electronic sensors. Fluorescence microscopy produces clear images of thin samples, such as brain sections or neurons growing in a dish.

The situation is a lot more challenging for imaging neurons in the intact, living brain. Because the brain's gray matter is so densely packed with cell bodies, neurites, supporting cells, and blood vessels, traditional microscopy falls short of capturing clear, useful images. The brain is as impenetrable to light as a glass of milk, and mostly for the same reasons: When visible light enters the brain, myriad refractile structures randomly deflect it, causing the images to lose contrast and appear hazy or distorted.

Fortunately, a different method, called two-photon microscopy, has provided a solution to the opacity problem. Using the weird physics of quantum mechanics and extremely intense illumination, two-photon microscopy produces crisp images, even in living tissue. The key difference with respect to standard microscopy is the type of illumination light. In two-photon microscopy, invisible infrared light is used to excite fluorescent molecules in the sample, whereas in standard microscopy visible light excites the same types of molecules. In both cases, the resulting excited molecules emit light that we can see. Because infrared photons have lower energy than visible photons, two infrared photons have to effectively combine into one to be energetic enough to excite the fluorescent molecule, resulting in "two-photon" excitation of fluorescence (Figure 1). This strange phenomenon was predicted 85 years ago by the German physicist Maria Goeppert in her doctoral thesis.[1]

Because the process of two-photon excitation of fluorescence is extremely inefficient, producing images for microscopy in reasonable time requires illumination light more intense than that on the surface of the sun. A special property of lasers is that they can be focused to a uniquely tiny spot, boosting the concentration of photons, or light intensity. For this reason, two-photon excitation was observed experimentally only after the laser was invented in 1961.

Despite the groundbreaking invention of the laser, standard laser light still was not sufficiently bright for two-photon microscopy. It took about another 20 years to develop specialized lasers that produce trains of extremely short pulses of light (one-tenth of a trillionth of a second). These potent laser beams effectively concentrate light in time and, at the point of focus of the two-photon microscope, produce a brightness that exceeds all other light intensities in the known universe.

Aside from the cutting-edge lasers, the remaining mechanics of two-photon microscopy are fairly simple. The laser beam is shaped to a tiny point and scanned in three dimensions over the specimen of interest. Then the fluorescence emitted by the sample is collected as a signal. This fluorescent signal, which represents a single point of the sample illuminated by the laser, is converted into a pixel, showing the corresponding location in the image. Thus the final, complete image is built one laser focal point and corresponding pixel at a time. Although it sounds painstaking, the lasers can move over the brain at high speeds, like a laser light show, producing images at rapid rates, comparable to a motion picture.

Where It Began

Two-photon microscopy was born more than two decades ago at Cornell University, where experts in microscopy and laser physics worked under one roof.[2] But before it took off, many scientists were skeptical about its use in biology, surmising that the kind of extremely intense light two-photon microscopy required was unsuitable to image living biological specimen. On top of that, the high cost and complexity of the early lasers presented a formidable obstacle to biologists. Most importantly, biologists had yet to find a standout application for two-photon microscopy.

The skepticism evaporated with developments in the mid-1990s at Bell Laboratories. Winfried Denk, one of the inventors of two-photon microscopy, and his collaborators noticed they could actually use the technique to image thick tissues while preserving the wonderful properties of fluorescence microscopy: high spatial resolution and high contrast. Two-photon microscopy is advantageous for imaging thick tissue for multiple reasons[3]. First, the infrared excitation light travels much farther in tissue without having its path perturbed, so the tiny laser focus can be maintained much deeper for imaging. Second, if tissue deflects a photon, it simply bounces out of the brain without producing fluorescent background noise. Third, fluorescence photons can be gathered efficiently to produce a bright signal. And, as it turns out—rather remarkably—the intense infrared excitation light is harmless to biological tissue. As a result, two-photon microscopy produces spectacularly detailed images of living neurons in intact nervous systems (Figure 2).

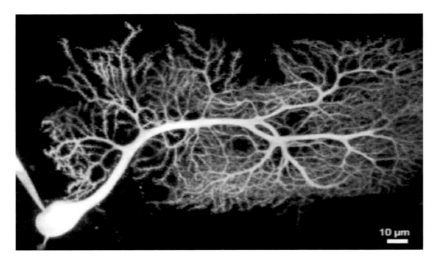

10 μm

Figure 2. Purkinje neuron in a brain slice imaged with two-photon microscopy. A fluorescent dye was injected into the neuron through a glass micropipette (left). Note the micrometer-size spines on the tree-like dendritic arbor of the neuron. (Image from Denk & Svoboda, Neuron, 1997)

Lighting Up the Brain

By itself the brain is nonfluorescent and appears black when viewed under a fluorescence microscope. Visualizing neurons with a two-photon microscope requires that fluorescent substances be added to the brain. Biologists had figured out how to make many distinct parts of neurons fluorescent, for example using antibodies with attached fluorescent molecules, but these methods were not usable in the intact brain. At Bell Laboratories in 1997, Denk, David Tank, David Kleinfeld and I produced the first clear images of neurons in the intact brain using two-photon microscopy. We threaded a tiny glass tube into the brain and injected fluorescent dye into the cytoplasm of a single neuron (similar to Figure 2). Trapped inside the neuron, the dye spread throughout the cell's axons and dendrites, making it light up like a Christmas tree. Using two-photon microscopy, we saw the neuron in its native habitat in exquisite detail.

But injecting one neuron at a time in an intact brain is a difficult and tedious procedure, and it often disrupts the targeted neuron. Introducing fluorescent substances to living neurons in the intact brain in an efficient and nondestructive manner created a new technical challenge. An unexpected solution emerged from the sea when scientists discovered a naturally occurring fluorescent protein in crystal jellies, corals, and anemones.[4] Osamu Shimomura, Martin Chalfie, and Roger Y. Tsien received a Nobel Prize in Chemistry in 2008 for their "discovery and development of the green fluorescent protein (GFP)." These proteins are fully encoded by DNA sequences, meaning that they can be manipulated using the tools of molecular biology. Scientists now modify the genomes of neurons, or infect neurons with engineered viruses, to coax them into producing fluorescent proteins. Although GFP is the most famous of the fluorescent protein family, others can glow blue, orange, or red. The combination of two-photon microscopy and fluorescent proteins has made imaging in the intact brain routine.

Imaging Neuronal Structure

Since the early days of brain research, scientists thought that neurons changed structurally during long-term memory formation. While experiments in invertebrate nervous systems suggested that simple memories are encoded in the growth of new synaptic connections between neurons, no evidence suggested the same for complex mammals, such as humans. Imaging tools weren't sophisticated enough to test the hypothesized changes. In 2002, Brian Chen, Josh Trachtenberg, and I, then working at Cold Spring Harbor Laboratory, decided to image the same neurons in the brains of mice for a period of months.[5] We used mice with modified genomes in which one-hundredth of 1 percent of neurons expressed GFP. The labeled neurons clearly stand out in two-photon microscopy, with virtually no background from the unlabeled neurons. The images were sufficiently crisp that we were able to track changes in even the smallest neuronal structures (Figure 3).

The dendrites of most neurons, for example, are studded with tiny protrusions called spines, about as long as one-hundredth of the width of a human hair. Spines harbor the synaptic inputs to neurons and bridge the gap across neurons, from dendrites to axons. A series of time-lapse images captured over days and weeks revealed that a subset of the spines appear and disappear, connecting and disconnecting neurons. During the experiment,

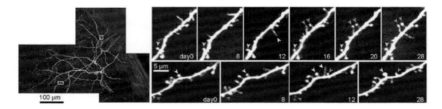

Figure 3. Long-term imaging of neuronal structure in the intact brain. Left, view of the dendritic arbor of a single neuron expressing green fluorescent protein. The fluorescent striations on the right are the surface of the brain. Yellow boxes correspond to the zoom-in on the right. Right, two high-magnification time-lapse sequences showing structural changes at the level of dendritic spines. Yellow arrows indicate persistent spines, orange indicate new spines, and green indicate lost spines. (Data from Holtmaat et al., Nature, 2006).

we would change the experience of the animals through behavioral training; for example, mice had to use their whiskers to distinguish, progressively, differences in locations of an object. We saw the rate of new connections increase as the mice learned new tasks, but only in the parts of the brain relevant to learning and with a time course similar to learning. What's more, mutant mice with learning deficits showed reduced ability to make new connections. These experiments, and many others like them, have provided support for the hypothesis that structural changes in neural networks support information storage in the brain.[6]

These long-term imaging experiments revealed structural changes at the level of synapses, but from a broader perspective, they also indicated remarkable structural stability. For example, the dendrites on which the spines sit persist over the entire life of the animal; this is also largely true for the axons. It's likely that, in the dense mesh of intermingled axons and dendrites, those nearby one another are neighbors for life. Between neighbors, though, these synapses fluctuate between formation and elimination to support the making of new memories.

Imaging Neural Activity

In the neuronal symphony, each neuron's tune is made up of sequences of electrical signals called spikes, which represent information. Through synaptic communication, these spikes influence the pattern of spikes in other neurons. Each neuron produces a characteristic pattern of spikes, like each instrument in a gigantic symphony orchestra. The flow of spikes through large collections of connected neurons underlies the processing of information in the brain. Imaging the structure of neurons alone is a bit like analyzing the shapes of the instruments in a symphony orchestra: suggestive and informative, but not sufficient to perceive the music. Recently it has become possible to use two-photon microscopy, together with fluorescent probes of calcium, to measure spikes in large numbers of neurons simultaneously and literally produce movies of the neural symphony in action.

Neurons have specialized mechanisms to keep their calcium levels very low, and, for reasons we do not yet understand, neuronal spikes cause rapid and large calcium changes in neurons. Scientists find calcium, a ubiquitous intracellular signal in most cells, an alluring target to monitor. Shortly after accomplishing fluorescent protein cloning, protein engineers started tinkering with DNA sequences to make fluorescent proteins that change in response to calcium binding. In the past decade, a variety of calcium-reporting proteins have been iteratively fine-tuned in efforts to image the spikes found in neural populations. The GENIE project at Howard Hughes Medical Institute's Janelia Research Campus, a highly collaborative project involving multiple investigators, solely strives to engineer the next generation of genetically encoded indicators for neuronal function. The GENIE project's "GCaMP6" proteins, which consist of GFP fused with the calcium-binding protein calmodulin, have proved particularly useful.[7] Upon binding to calcium, GCaMP6 becomes 50-fold brighter. Quiescent neurons expressing GCaMP6 are essentially black but produce bright flashes of green fluorescence immediately after a spike. Using two-photon microscopy, researchers can readily image these temporal changes in fluorescence and interpret them in terms of spike trains (Figure 4). GCaMP6 is so sensitive that it can track activity in thousands of neurons when imaged using two-photon microscopy.[8]

In the past decade, imaging neural activity has progressed into a critical method in brain research. GCaMP6 is used by more than 1,000 laboratories worldwide to study fundamental questions in neuroscience, but also to understand mechanisms underlying neurodevelopmental and neurodegenerative disorders. Now, imaging large populations of neurons while an-

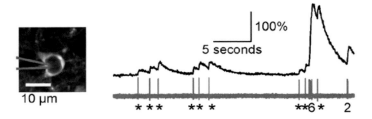

Figure 4. Left, image of the neurons with schematic of a pipette (red) that is used to directly record spikes. Right, comparison of fluorescence changes (top) and spike trains (bottom) for the neuron. Numbers below the spike train indicate the number of spikes in a burst; asterisks indicate single spikes (data from Chen, Wardill, et al., Nature, *2013).*

imals perform behavioral tasks is routine. Researchers are even combining two-photon imaging simultaneously in multiple brain regions with virtual reality, in which mice might navigate a visual or tactile environment in search of a reward. Through statistical analysis of activity patterns in neurons, movies of GCaMP6-expressing nerve cells are beginning to reveal the relationships between the behavior and activity of specific neurons. In this way, scientists are learning how spike trains represent information in the brain and how these representations change with learning.

What's Ahead

Brain research, which relies on progress in computation, electron, and light microscopy, as well as DNA and RNA sequencing, molecular biology, genetics, chemistry, and applied physics, is entering a rapid phase of discovery. Research methods are evolving and combining in interesting ways to boost brain research. Rapid sequencing methods developed for cancer biology now measure gene expression in single neurons and are beginning to provide a catalog of the brain's neuron types.[9] New anatomical methods derived from virology enable a mapping of the connections between those neuron types.[10] Optogenetic reagents allow scientists to manipulate the electrical signals in neurons with light and evaluate how these neurons contribute to animal behavior.[11]

The story of two-photon microscopy illuminates several points with respect to technology development and scientific discovery. New technologies ring in rapid phases of discovery. Often these technologies bring together multiple disciplines in unexpected ways. Two-photon microscopy would have struggled to take off without the boost of laser technology. And without the cloning and engineering of GFP, imaging the intact brain would have remained an artisanal pursuit of interest to relatively few biologists. Finally, so-called incremental improvements to technology can open up new vistas and shape science just as much as completely new advances. Multiple rounds of quantitative improvements to GFP-based sensors ultimately made the once unattainable holy grail of brain research—images of neural activity in the living brain—possible in many laboratories throughout the world.

5

A New Approach for Epilepsy

By Raymond Dingledine, Ph.D., and Bjørnar Hassel, M.D., Ph.D.

Raymond Dingledine, Ph.D., is professor and chair of the Department of Pharmacology at Emory University School of Medicine. Dingledine received his Ph.D. in pharmacology under Avram Goldstein at Stanford and postdoctoral training from Leslie Iversen and John Kelly at Cambridge, UK, then Per Andersen at Oslo. His research focuses on the pharmacology of glutamate receptors and on the causes of epilepsy. He serves on the scientific advisory boards of the Epilepsy Project and NeurOp, a biotech start-up that he co-founded. He is also chair of the investment committee of the Society for Neuroscience and a member of the National Academy of Medicine.

Bjørnar Hassel, M.D., Ph.D., is a neurologist and a biochemist. He is professor and senior consultant at the University of Oslo, Department of Neurohabilitation. The department serves adults with developmental disabilities, both intellectual and physical. Hassel works with persons who have severe cerebral palsy or profound intellectual disability and autism, and who are restricted in their ability to communicate. He explores the use of sensors for physiological parameters (pulse, plasma glucose, etc.) as a means by which they can communicate their needs and their degree of well-being. His preclinical research centers on cerebral energy metabolism, including how it is affected by antiepileptic drugs.

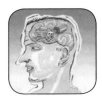 *About one-third of the 65 million people worldwide affected by epilepsy are treatment-resistant, and the degree to which they suffer from seizures and convulsions can vary widely. Problems occur when nerve cells in the brain fail to communicate properly. A new study has found that inhibiting an enzyme that is critical in metabolic communication has an antiseizure effect in epileptic mice. These findings, the authors believe, may very well initiate a shift to new therapeutic approaches.*

IMAGINE A DOCTOR TELLING YOU THAT you have to change your diet to one with few carbohydrates in favor of high-fat cheeses, butter-fried steaks, bacon and eggs, and eggnog—all while you snack in between meals on macadamia nuts. Sounds great initially, but most of us would tolerate only a few days of eating this way. Yet many young children with epilepsy, who do not respond to conventional medications, benefit from just such a diet. Strict adherence to the so-called ketogenic diet (i.e., with minimal calories from carbohydrates) can often reduce their seizures enough to allow them to attend school and experience the joys of growing up. The diet was developed to mimic the effects of fasting, which has been known since antiquity to afford some seizure control.

Epilepsy and epileptic seizures affect nearly three million Americans and 65 million people of all ages around the world. According to the International League Against Epilepsy, seizures and epilepsy are not the same: "An epileptic seizure is a transient occurrence of signs and/or symptoms due to abnormal excessive or synchronous nerve cell activity in the brain. Epilepsy is a disease characterized by an enduring predisposition to generate epileptic seizures and by the neurobiological, cognitive, psychological, and social consequences of this condition. Translation: a seizure is an event, and epilepsy is the disease involving recurrent unprovoked seizures."

In fact, "epilepsies" are a group of neurologic disorders. When one or more neural circuits in the brain develop a chronically low seizure threshold, normally innocuous stimuli (external to or within the brain) can trigger a group of nerve cells to fire at once. This abnormal synchronicity is a seizure.

The disease has been known for ages. A very early reference is found on a Babylonian tablet in London's British Museum dating from approximately 1060 BC, which refers to "the falling disease," with the subjective aura (an ominous feeling) and the subsequent seizures themselves ascribed to the work of childless demons who viewed humans with envy and spite.[1] Hippocrates argued around 400 BC that epilepsy is a physical disorder of the brain, but he was widely disbelieved.[2] Over the next 2,000 years, seizures were treated by bleeding, exorcism, trepanation (a hole is bored in the skull), and ingestion of silver nitrate or bromides.

Over the past several decades more than 30 anticonvulsant medications have been developed. They pass via the bloodstream into the brain and dampen seizures by reducing the excitability of brain cells. They act on a restricted number of molecular targets in the brain. Some of the drugs act on ion "channels" that allow sodium, calcium, and potassium ions to pass into and out of brain cells. Others potentiate the major inhibitory system of the brain, which uses the neurotransmitter GABA to dampen nerve cell excitability. Additionally, there are drugs that act on the synaptic vesicle protein SV2A, and the AMPA subtype of glutamate receptor.

The ketogenic diet (KD) was introduced in the 1920s in between the first two modern antiseizure medications, phenobarbital in 1912 and phenytoin in 1938. Even though some patients did not respond to these drugs and improved with the ketogenic diet, it nonetheless fell into obscurity as more antiseizure medications were introduced. Then, as now, physicians found it far easier to prescribe a pill than to teach their patients that all that was required was to adhere to a rigid and restricted eating regimen. But even with the plethora of anticonvulsant drugs that are now available for people who live with epilepsy in the developed world, a full one-third of epilepsy patients still do not respond to any medication. This situation led to the creation of a childhood epilepsy center and seizure clinic in the mid-1970s at the Johns Hopkins Department of Neurology/Neurosurgery Hospital Clinic, where the ketogenic diet was resurrected as an option.

A drug that works for everyone, though, remains the goal. Because no magic pill exists to eliminate epilepsy as such, the search for new antiseizure medications has continued, especially studies of drugs that have novel molecular targets in the brain.

Hope on the Horizon

Last year a study reporting an unexpected molecular target that could spawn a new generation of anticonvulsant drugs stirred great interest.[3] Sada et al. reported the results from four experiments. The researchers started by looking at the effect in brain slices of bathing nerve cells in a solution that contained beta-hydroxybutyrate (BHB) rather than glucose (sugar) as an energy source. BHB is made in the liver when the body breaks down fat, rather than carbohydrates, for energy. This switch from carbohydrate to fat metabolism is, in fact, what occurs in people when they fast or when they are on the ketogenic diet. The study found that BHB *hyperpolarized* nerve cells, that is, rendered them less excitable and more stable, and thus less prone to epileptic activity.

When on the ketogenic diet, blood levels of sugar decrease while blood levels of BHB increase to exert its stabilizing influence. Even so, blood glucose remains around one-half of the usual level.[4,5] When BHB was added to a bathing solution that contained a little rather than no glucose, the cells did not hyperpolarize. Rather, they remained active and prone to epileptic activity. This finding suggests that glucose might offset the stabilizing effects of BHB.

Returning to the nerve cells in Sada's experiment, the key question is whether they became more stable from the presence of BHB or the absence of glucose. This question gets at the heart of how the ketogenic diet works, because it is still not clear whether its antiepileptic effect is due to low blood levels of glucose or to the high levels of BHB that occur when the body breaks down fat for energy.[5,6]

Sada et al. investigated this issue by asking whether the glucose-to-BHB switch acts by reducing the formation of pyruvate and lactate. These organic compounds are produced when glucose is broken down, and each can be converted into the other (interconverted) by the enzyme lactate dehydrogenase (LDH). Through a series of experiments, the researchers found that: 1) lactate could indeed undo the stabilizing effect of BHB by reversing its hyperpolarizing action; 2) lactate caused this effect after it had been converted into pyruvate by LDH, because 3) inhibition of LDH by the small molecule, oxamate, eliminated the anti-hyperpolarizing (depolarizing) effect of lactate, but not that of pyruvate, which continued to depolarize nerve cells even when LDH was inhibited by oxamate, 4) just exposing the nerve cells to oxamate caused them to become hyperpolarized.

At a Crossroads

This led to a crucial question: Does pyruvate reverse the hyperpolarizing (stabilizing) effect of oxamate by acting as a nutrient—or does it do something else?

Pyruvate is a "keto" acid, carrying a keto (C=O) group. Sada et al. showed that other keto acids (alpha-ketobutyrate and oxaloacetate, see Figure 1) also reversed the stabilizing effect of oxamate, but that ATP or other energy metabolites derived from pyruvate were ineffective. Together, these findings suggest that the depolarizing effect of pyruvate is related to its ability to scavenge the coenzyme NADH (Figure 1) rather than its role as a nutrient. It is likely that BHB, a hydroxyacid, becomes converted to the ketoacid "acetoacetate" with an accompanying formation of NADH. This is especially likely to be the case when BHB is delivered in large quantities, as in a solution that bathes the nerve cells.

Mechanisms to Consider

Thus, the stabilizing or hyperpolarizing effect of BHB appears not to be related to energy production in the nerve cells, but instead could be caused by its promotion of an NADH-dependent process regulating the excitability of nerve cells. Several candidate mechanisms can be considered to explain how LDH inhibition hyperpolarizes nerve cells; much work remains to pin down the exact series of events.

The next question that Sada et al. asked was how pyruvate, converted from lactate, ends up in nerve cells. They found that it probably is delivered to the nerve cells by another cell type, astrocytes. When glucose enters the brain from circulation, it is partly taken up into astrocytes. Inside these cells, some of the glucose will be converted into lactate, which leaves the astro-

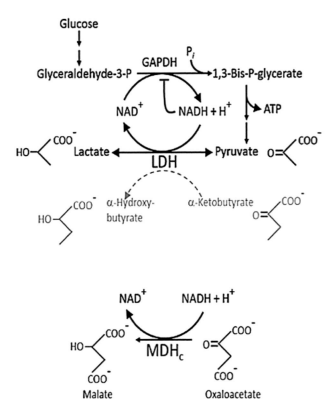

Figure 1. A simplified scheme of the glycolytic pathway, emphasizing the generation of NADH and the purpose of LDH in consuming excess NADH. In the cytosol NADH is formed in the glyceraldehyde-3-phosphate dehydrogenase (GAPDH) reaction. GAPDH is strongly inhibited if NADH accumulates. GAPDH catalyzes the oxidation and phosphorylation of glyceraldehyde-3-phosphate by inorganic phosphate (Pi), producing 1,3-bis-P-glycerate. In the subsequent enzymatic step the phosphate group is donated to ADP to produce ATP. After another three enzymatic steps, pyruvate is formed. The conversion of pyruvate to lactate, catalyzed by LDH, consumes NADH and relieves GAPDH of its inhibition by NADH, allowing the cycle to continue. α-Ketobutyrate (in blue), which was used by Sada et al., may substitute for pyruvate in the LDH reaction. Oxaloacetate (purple) may also consume cytosolic NADH in the cytosolic malate dehydrogenase reaction (MDHc), in which oxaloacetate is converted to malate at the expense of NADH.

If this system is flooded with lactate (e.g., by lactate exported from astrocytes to nerve cells), then NADH would accumulate. If the system is flooded by pyruvate (e.g., by experimental injection of pyruvate into nerve cells), then NADH levels would be reduced.

cytes. Nerve cells then take up the lactate from astrocytes through a specific transport protein, completing an astrocyte-to-neuron lactate shuttle.[7-9] The Sada group concluded that lactate from astrocytes is an important source of "epileptogenic" (seizure-promoting) pyruvate in nerve cells.

Sada and colleagues had found that bathing nerve cells in the LDH inhibitor oxamate caused the cells to hyperpolarize and become electrically stable. They followed up this finding by injecting oxamate into the hippocampus of mice that had been rendered epileptic after treatment with kainate (a neuroexcitatory amino acid). Oxamate reduced seizures in this mouse model of epilepsy, suggesting that LDH inhibition could be antiepileptic in vivo too. Now that LDH was identified as a potential target for antiepileptic therapy, Sada et al. looked for antiepileptic drugs (AEDs) that might inhibit LDH. The researchers screened 20 AEDs and found that a little-used AED, stiripentol, did inhibit LDH. Stiripentol is currently used to treat a rare and devastating epileptic condition known as Dravet syndrome or severe myoclonic epilepsy of infancy. They then tested similar molecules for LDH inhibition and found that the structurally simpler compound, isosafrole, was a more potent inhibitor of LDH than stiripentol itself.

Pyruvate depolarizing nerve cells

The picture emerging from the work of Sada et al. is that pyruvate facilitates epileptic activity by depolarizing nerve cells. It had been shown earlier that rapid injection of pyruvate could actually cause seizures.[10] Sada and colleagues showed that blocking the enzyme responsible for pyruvate formation from lactate (LDH) has an anti-seizure effect.

This remarkable study helps explain the well-known observation that eating a sugar-laden cookie can quickly promote seizures in a child on the ketogenic diet:[11] when the sugar (glucose) from the cookie is metabolized, it forms pyruvate; the pyruvate then rapidly reverses the stabilizing effect of BHB. Further, the Sada group identified the role of LDH in the glucose effect, and they showed that pharmacological inhibition of LDH exerts marked anti-seizure effects in epilepsy animal models—thus fingering LDH as the first new target for discovering epilepsy drugs since 2004.[12] Important for understanding their results is the recognition that the seizure-promoting mechanism of the glucose metabolites—pyruvate and oxaloacetate—is unrelated to ATP generation but could involve NADH scavenging.

Key steps remain in the quest to develop novel antiseizure drugs based on reduced formation of pyruvate. First, we might identify additional antiseizure targets by identifying the mechanism involved in oxaloacetate-induced depolarization. A good next step for this would be systematic testing of other Kreb's cycle metabolites, especially NADH, for their ability to reverse oxamate-induced hyperpolarization. Second, there are multiple, cell-specific but functionally similar forms of LDH; further drug screening might allow glycolytic inhibition that is restricted to nerve cells, which could reduce side effects. Third, their unexpected finding that inhibitory interneurons are not hyperpolarized (stabilized) by LDH inhibitors, although fortunate for epilepsy therapy, is worth following up.

Finally, inhibition of glucose metabolism to treat epilepsy has been reported previously, using 2-deoxyglucose (2-DG),[13] and it would be interesting to determine if 2-DG and isosafrole block one another's action or are synergistic. The answer could inform whether reduced glycolysis or some unrelated action is responsible for the antiseizure effect of 2-DG. If the two are synergistic, combination therapy could be warranted. While not quite "epilepsy diet in a pill," the study by Sada et al. points the way to a "starvation in a pill" strategy— i.e., pharmacologically simulating fasting— for seizure control. Antiseizure drugs have been repeatedly repurposed for other neuropsychiatric disorders such as bipolar disorder and neuropathic pain. If chronic treatment with 2-DG and an LDH inhibitor prove safe in people, might we add another repurposed clinical use to this list—weight control?

6

The Neuro Funding
Roller Coaster

By Harry M. Tracy, Ph.D.

Harry M. Tracy, Ph.D., is the founder and president of NI Research in Cardiff, CA. NIR's bimonthly publication, *NeuroPerspective*, is utilized by pharmaceutical companies and venture capital professionals. NIR has also published *NeuroLicensing* and the *Private CNS Company Review*. NIR provides consulting services to pharmaceutical companies and to venture capital/private equity groups. Tracy also practiced for 30 years as a clinician and consultant in a variety of psychiatric and neurological settings. He received his Ph.D. from the University of Miami and completed his clinical training at Massachusetts General Hospital/Harvard Medical School. He was formerly a research associate in the Department of Neurology at the University of California, Davis.

Compared to the money dedicated to cancer and cardiology, funding for neuroscience research has lagged behind for decades. But things are starting to change. From the White House's Brain Initiative to the Ice Bucket Challenge for ALS to some recent sizable gifts to universities, money for brain research appears to be on the rise. But, as our author explains, research and development funding from private and corporate lenders for cognitive neuroscience—an area that he has spent years tracking—is also vital to the quality of life for millions of people.

AS MUCH AS WE LIKE TO FRAME RESEARCH for neurology and psychiatry as being rooted in harnessing science to improve the practice of medicine, the arguably crass but sobering reality is that while applied neuroscience may be ultimately judged by its scientific/medical achievements, the process depends upon the availability of money for the arduous research and development journey. The most extreme roller coasters found at amusement parks have nothing on the stomach-churning oscillations of funding for research devoted to neurology and psychiatry—other than the latter is timed in months and years, rather than seconds. The past decade has seen the flow of resources for applied neuroscience sink to a stunning nadir, followed by an even more astounding resurgence.

But this recovery has been very uneven, and some areas, particularly psychiatry, have yet to rebound to anything near their former levels of fiscal well-being. Even this retrospective appraisal comes during a period of flux reflective of, and exacerbated by, geopolitical, political, and demographic dynamics currently at a peak. We recently experienced the best year for neurotherapeutics funding in over a decade, perhaps ever. But as early 2016 unfolded, the gains of 2015 receded like a favorite vacation spot shrinking in the rearview mirror; there has been a dramatic retreat from investment by both institutions and pharma partners.

As of the end of April, the annualized projection for institutional funding/investment was down more than 50 percent from last year's total, and partnering (in terms of disclosed upfront payments) was down more than 75 percent. It remains to be seen to what degree this is a transitional hitch in the recovery process or a return to a painfully familiar climate of angst and parsimony. We believe it is the former, but this is a hypothesis yet to be tested by time, and there is more than ample pessimism about resource availability to be found in the venture-capitalist community.

With that caveat noted, we will examine the past decade in the funding of neuropsychopharmacology, as well as cell and gene therapeutics, in terms of the enormous perturbations that have occurred within the institutional investment climate for the neuro sector and in the pharma industry's willingness to partner neurology/psychiatry programs in development by smaller firms. We will focus strictly upon these two fiscal domains. The third major research and development resource, governmental grant funding, took on greater salience during the recent period of fiscal deprivation, but its role is to some degree compensatory: Such funding partly (but never entirely) makes up for shortfalls in investor/pharma dedication to neurotherapeutics.

Serendipity and Stasis

Our discussion will not address the massive scale of unmet medical need to be found in populations suffering from neurological and psychiatric illness. Nor will we analyze the lack of substantive change in treatment options for these patients. Suffice it to say that, in spite of the billions of dollars spent searching for better treatments for neurological and psychiatric disorders, and the myriad advances made in basic neuroscience, when it comes to real world therapeutic drug options, the situation has been one of near stasis. Our antidepressants do not differ significantly from those that were available 20 years ago; the same can be said of our antipsychotic options, used primarily for schizophrenia and bipolar disorder; and for the modest cognitive enhancers marketed for Alzheimer's.

The one exception—an area where significant advances have occurred in terms of efficacy and ease-of-use—is in the treatment of Relapsing-Remitting Multiple Sclerosis (RRMS). The beta-interferons gave rise to IV natalizumab, then the oral S1P targeting compounds, and the oral fumarate-based drugs. It is the one disorder where we are now able to actually slow the progression of the disease, rather than simply mask or suppress symptoms, and the repertoire comprises a range of choices with a spectrum of risk-reward profiles from which patients and their physicians can choose. It is also the area where the regeneration of what has been lost, in terms of neural circuits and functioning, is now being first attempted. Advances for RRMS represent the way we hope neurotherapeutics for other disorders will evolve over the next two decades.

There are several reasons for this era of stasis. First, the great psychiatric drug classes that emerged during the 1950s and 1960s were largely the product of serendipity; the clinical observation of therapeutic effects that led to post-hoc hypotheses explaining how these drugs might work were—for the most part—questionable at best, wrong at worst. They failed to provide a road map for their successors; many scientists embarked on research journeys launched by assumptions that turned out to be incorrect and guided by processes that misinformed. There was a false sense of confidence based on the commercial success of new drug classes that had become popular due to a better side-effect profile, rather than improved efficacy, like the SSRIs and second-generation antipsychotics. The pharmaceutical industry became, for lack of a better word, "lazy" when it came to internal R&D; imitation was frequently more prized than discovery, as many companies tried to piggyback on the success of earlier drugs through tweaking rather than innovating. To the degree to which innovation was permitted and funded, there was a tendency toward premature closure, choosing new mechanisms for full development without adequately auditioning the range of alternatives. The single best example of this is Alzheimer's, where the bulk of research funding and testing over the past 20 years has relied upon an amyloid hypothesis that, even now, has yet to prove itself to be valid.

A Crisis of Faith

Because the brain is often described as the most complex structure in the known universe, neuroscience is fundamentally more challenging and less advanced than other areas of medicine. One key problem has been—and continues to be—that the tools with which neuroscience R&D is carried out have been inadequate to the task, and generally less effective in their application than those available to other therapeutic endeavors. This has yielded an inevitable string of failures.

In the world of animal models, the theoretical underpinnings have mostly been dubious, and the issues numerous. The system used for classifying disorders is based on categories that date back a century or more. What's more, the pathophysiological roots of most disorders are unknown; targets for intervention have been generally based on theories derived from animal models of ambiguous relevance, and are located behind the blood-brain-barrier, making it difficult to get drug candidates where they were needed. It has been for the most part impossible to be sure if target engagement has been achieved. And the endpoints by which clinical status and progress are measured in human testing still tend to be ambiguous and subjective, particularly in psychiatry.

From a pragmatist's point of view, the question might better boil down to: Why would anyone invest in this area? The customary and admittedly true answer generally involves "unmet medical need," but the existence of such needs, great as they may be, does not in itself form a bridge to the treatments being developed for them. To a large extent, we have been flying blind, without much in the way of instrument assistance. The definition of success has been binary, determined by whether the flight landed, or ended in a crash. The wreckage of many highly touted programs litters the runways of the biopharma industry and has come to dominate the perception of the neurotherapeutics area in the eyes of many (albeit not all) investors and pharma companies as being too risky.

Confidence in biopharma's neuroskillset was gradually ground down to a nub via a drawn-out series of high-profile failures, which led investors to question whether neuroscience had any idea what it was doing. The full list of failures is too long and disheartening to review in detail, but clinical landmarks that played an important role in squandering industry credibility include:

- Myotrophin and ALS (1997)
- Substance P and depression (1999)
- Free radicals and stroke (2006)

- Bapineuzumab and Alzheimer's (2012)
- Pomaglumetad Methionil and schizophrenia (2013)

The net effect of these failures over a period of 15 years was to flag neurology and psychiatry as "too hard," leading to the nearly complete departure of GlaxoSmithKline from neuroscience and to significant contractions of neuroscience programming at Sanofi, Merck, Lilly, and others.

No Exit: The Existential Angst of the Venture Capitalist

Even as the success rate of neuroscience R&D plummeted, the First World was approaching a macroeconomic near-death experience. The fiscal crisis of 2008-10 led to a dramatic retraction of capital investment, and investors became highly skittish when it came to investment risk. This made it extremely difficult for privately held biotech companies to go public, and threw a major obstacle into the cycle by which capital enters and exits the biopharma system.

While travelers tend to ignore the safety feature demonstrations provided by a flight attendant before each takeoff, assuming that exits available for emergencies will not be needed, such is not the case for venture capital and institutional investors, for whom the location and timing of an exit are core components of their investment model. In biopharma, such exits take the form of IPOs, wherein public money comes into a company, replacing much of the private investment that sustained the company to that point, and provides continued liquidity via stock sales. The simultaneous withdrawal of Big Pharma as a potential acquirer, along with this closing of the IPO "window," meant one thing above all for anyone contemplating investment in the neurotherapeutics area: Once invested in a small company, there was no near-term exit at hand.

Even worse, with no promise of new investors, the initial investors were increasingly confronted by the choice of either doubling down on a high-risk venture or letting it wither and die. Venture capitalists, who operate in predefined life -cycles, typically less than ten years, could not credibly assure investors that they would be able to retrieve their investment, hopefully with profits attached, in that promised timeframe. This, in turn, reduced the inflow of resources to these venture capitalists, impairing their ability to sustain old investments, let alone make new ones, and thus produced a vicious cycle of the first order.

To illustrate, consider the casualty rate among 81 private cognitive neuroscience companies that NI Research tracked, beginning in 2003: Of those companies, only 19.7 percent provided an exit for their investors (12.3 percent via acquisition, 7.4 percent via IPO), and 61 percent went out of business entirely, representing a complete loss for their investors. The remaining 19 percent have continued in private operation, many of them barely alive, a herd of neurotech zombies. They are far more likely to end up in the failure than the success column.

Venture Capital Entrances and Exits: IPOs 2004-15

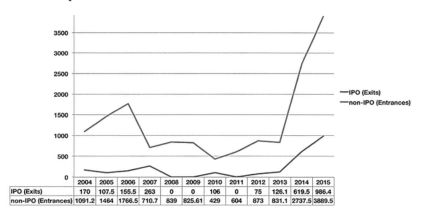

	2004	2005	2006	2007	2008	2009	2010	2011	2012	2013	2014	2015
IPO (Exits)	170	107.5	155.5	263	0	0	106	0	75	126.1	619.5	986.4
non-IPO (Entrances)	1091.2	1464	1766.5	710.7	839	825.61	429	604	873	831.1	2737.5	3889.5

This chart illustrates the correlation between the availability of IPO exits and the availability of investment for new and young central nervous system (CNS) companies, which represents the entrance of money into the small company CNS arena. Over the ten years from 2004 through 2013, there were a total of just 21 IPOs by private CNS companies. In fact, there were none at all in 2008-09. But in 2014-15, there were 23 CNS companies that completed IPOs. As the number and total raised by these IPOs began to accelerate, there was an even more dramatic flow of investment into the CNS sector, far exceeding the amount that exited. Reinforcing this transformation was the less dramatic, but still vitally important, return of some large and midsize pharma companies to neuroscience, while partnering external research. Indeed, many large companies have essentially outsourced much of their neuroscience research to smaller companies.

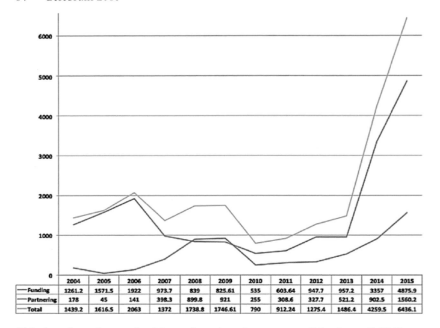

	2004	2005	2006	2007	2008	2009	2010	2011	2012	2013	2014	2015
Funding	1261.2	1571.5	1922	973.7	839	825.61	535	603.64	947.7	957.2	3357	4875.9
Partnering	178	45	141	398.3	899.8	921	255	308.6	327.7	521.2	902.5	1560.2
Total	1439.2	1616.5	2063	1372	1738.8	1746.61	790	912.24	1275.4	1486.4	4259.5	6436.1

This chart shows the past decade's transformation of resources overall for the small CNS company world, from both investors and pharma partners. The spectacular escalation of resource availability, particularly from investors, began toward the end of 2013 and accelerated over the next two years. What accounted for this turnaround? To some degree it was the economic recovery overall, particularly in the US, yielding capital that was looking "for a home." In spite of the legacy of failure and frustration, there were factors that began to make the neurotherapeutics area palatable, even desirable, to investors.

There is the belief, perhaps more akin to faith, that the genomics advances that allowed the development of precision medicines in oncology may yet prove useful in neuroscience, the greatest untapped market of all. Beyond these nascent genomics advances were the maturation of neuroimaging (e.g. amyloid plaque, tau), cerebrospinal fluid biomarkers, changes in nomenclature (the National Institute of Mental Health research domain criteria initiative), and technologies allowing the tracking of trial and treatment compliance. In aggregate, these constitute the harbinger of an era in neuroscience wherein guesswork is being replaced with something more substantive.

We are better equipped to track target engagement—at least for some targets—and RRMS is now seen as the leading edge, rather than the sole outlier. Induced pluripotent stem cell models, modified via gene editing, offer a more face-valid screen for early drug development than the animal models upon which the sector has too long relied. Remote biosensors and big data analytics offer the prospect of being able to sift huge datasets for meaningful relationships—defining pathways where new and more impactful targets may exist—and the concept of brain and brain pathology as based on networks of circuits can be reified and tested via techniques like optogenetics. Parsing this list, in a highly oversimplified fashion, it can be argued that there were three events or dynamics that set the stage for this transformation: personalized oncology, hepatitis C treatment, and neuroimaging.

- Personalized Oncology: The human genome was sequenced in 2000, but it has taken a long time for the genomics revolution to make a tangible difference in the practice of medicine. We would argue that it is in the area of personalized medicine in the treatment of cancer that the highest profile gain from genomics has been realized: Rather than relying upon trial and error (and tradition) in offering patients various cocktails of chemotherapeutics, oncology began to parse cancer into subcategories based on genetic factors. This altered the climate of frustration and delay that had grown around genomics and raised the question: Where else might this work? When it comes to large-scale, heterogeneous populations of uncertain etiology, nothing exceeds the scale offered by the worlds of neurology and psychiatry, and the belief system within the investment world has begun to shift toward the anticipation that genomics will render these disorders more comprehensible and tractable.

- Hepatitis C: A disease that formerly could only be treated palliatively turned into a disease for which "magic bullets" can provide a cure. The fact that these treatments can be premium-priced and yet make pharmacoeconomic sense has also hit home for the investment community. In an era where treatment pricing had turned into a war of attrition between payors, generic manufacturers, and pharma companies, and like personalized treatments for cancer, this portended a time where a successful new product could be once again expected to produce outsized profits. No area had been hit harder by the advent of generics than neuroscience as a whole, and psychiatry in particular, and this provided a road map for returning some pricing control to the neuropharm world.

- Neuroimaging: As was mentioned earlier, the success of disease-modifiers in the treatment of Relapsing-Remitting Multiple Sclerosis for a very long time was the only example of substantive, rather than incremental, progress in the treatment of CNS disorders. What made RRMS different? One critical element was the availability of imaging technology that could provide an objective, empirical measurement of therapeutic impact on the rate at which the disease progressed. As gadolinium-enhanced imaging was for MS, the advent of florbetapir as a means of empirically quantifying amyloid plaque in Alzheimer's has served as a beacon of hope that biomarkers would begin to take neurotherapeutics out of its familiar morass of "squishy," subjective endpoints. The fact that amyloid plaque's utility as a biomarker has yet to be fully established has been less important than its role as the poster child for a new era of objective measures in neuroscience. Other imaging markers (e.g., tau) and a plethora of blood and CSF biomarkers have emerged, albeit yet-to-be-proved in their ability to focus and accelerate CNS drug development.

Other Factors Carrying Weight

While the process of testing candidate drugs in clinical trials remains a high-risk, high-anxiety endeavor for neuroscience, new tools offer greater assurance that participants in drug trials are real patients, not professional patients, and are actually taking the medications as they are supposed to. Patients who are not ill or who do not take the medication being evaluated, simply are not a valid template for testing the efficacy of such a drug. We will never know how many clinical trials have been ruined by noncompliant patients; one must wonder how many potentially useful drugs had the signal of their therapeutic impact obscured by a flawed clinical testing process. Clinical trial professionals have come to realize, sometimes at odds with the companies sponsoring the trials, that in this context, "speed kills." Companies have started to cooperate in flagging fake patients and are beginning to explore technologies (like a chip on a pill) that allow accurate monitoring of what a patient takes, and when.

And in an environment where generic drugs have become king, the pharma industry and its investors have finally come to recognize that redundancy no longer is remunerative, and that they will have to grapple with the risk, along with the potential reward, of novel mechanisms for intervention. As confidence in the tools has grown, so too has the willingness of both pharma partners and investors to bet on higher-innovation, higher-risk programs, because making significant inroads on the huge, unmet needs of neurology and psychiatry will require new and disruptive technologies.

Neurotherapeutics is, of course, not a unitary construct, and it is informative to consider where resources are flowing more specifically. The breakdown over the past decade, parsing the area into four categories (Neurodegenerative Disease Modifiers, Neurology Symptomatics, Psychiatry, Pain), yields the following in terms of funding therapeutic subareas from 2009 to 2015:

Playing Favorites: Where the Therapeutic Areas Rank

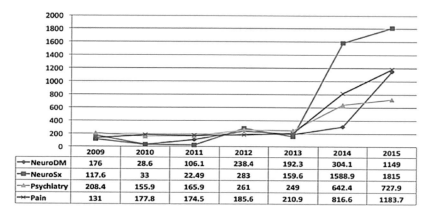

	2009	2010	2011	2012	2013	2014	2015
NeuroDM	176	28.6	106.1	238.4	192.3	304.1	1149
NeuroSx	117.6	33	22.49	283	159.6	1588.9	1815
Psychiatry	208.4	155.9	165.9	261	249	642.4	727.9
Pain	131	177.8	174.5	185.6	210.9	816.6	1183.7

After several years where no neuroscience area was favored in terms of funding, there was a dramatic surge in 2014, with the most spectacular rise in funding ocurring for symptomatic treatments for neurological disorders, including levodopa-induced dyskinesia, and cognition or psychosis associated with neurodegeneration. This is a relatively lower-risk area compared to programs aimed at slowing or stopping the course of a neurodegenerative disease, which received the least funding of the four in 2014. But in 2015, while there was continued, slowly growing interest in symptomatic treatments, investors finally became willing to fund disease modifiers, reflecting the belief that improvements in neuroscience tools would mean that these high-risk programs stood a better chance of success.

It should be noted that the willingness to put venture capital into higher-risk programs has been anything but across-the-board, but there are some venture capitalists (VCS) who have the long-term perspective and neuroscience background that allows them to be more risk-tolerant. Thus, in 2014-15, the roster of VCs leading investment rounds in highly innovative research included Fidelity Biosciences (Denali Therapeutics, Yumanity Therapeutics, and Forum Pharmaceuticals); Third Rock (Voyager Therapeutics); Atlas Ventures (Lysosomal Therapeutics and Rodin Therapeutics) and Clarus Ventures (Annexon). The fact that the Denali Series A round of $217 million was by far the largest such round ever completed by a CNS company provided a signal to more reticent VCs that smart money is starting to find its way into neurodegeneration research—and that they should consider participating, even if not leading. Another important development has been heightened activity from pharma companies investing through venture arms, giving them the benefit of the investment itself and insight into ongoing research activities, without having to buy in completely.

In the other therapeutic subdomains, Pain enjoyed steadily increasing investment, partly due to the growing visibility of opioid abuse, enhancing the potential prospects for novel analgesia alternatives. Finally, Psychiatry, long out of favor, rebounded somewhat in 2014, the most apparent trigger being the rising profile of the rapid-acting-antidepressant class, epitomized by ketamine, which for the first time in 20 years seemed to offer the potential of a genuinely differentiated new antidepressant option. But the investment in Psychiatry plateaued in 2015, its flat growth leaving it well behind the other three areas in garnering investment dollars.

No Place for Vertigo: The Oscillations Continue

Overall, the past decade or two have constituted a humbling process to which the neurotherapeutics sector has had to submit, where scientists have had to accept the limitations of their knowledge base and research tools, finally going back to the drawing board. This has led to the growing salience of genomics, biomarker analyses, brain imaging, and sophisticated behavioral assessment technologies, providing an entirely revised approach to drug target delineation and validation. To be clear, the confluence of these technologies has yet to come to fruition. Outside of multiple sclerosis, no new drugs have been identified, refined, and proved via the new generation of techniques.

But there is a renewed emphasis on empirical, scientific inquiry and validation, which has given the investment community hope that the brain will not continue to be a black box whose workings—and our impact upon them—can only be guessed at. Beyond imaging and biomarkers, there is an array of new tools that portend an era of greater productivity for neurotherapeutics, which inspired—at least through the end of 2015—a resurgence of optimism and an influx of resources. Whether that will continue will depend on macroeconomic forces completely beyond the control and influence of the biopharma industry, and the degree to which clinical successes begin to provide tangible proof that this is in fact a new era for neuroscience.

7

Drinking Water and the Developing Brain

By Ellen K. Silbergeld, Ph.D.

Ellen K. Silbergeld, holds a Ph.D. in environmental engineering from Johns Hopkins University, where she is a professor in epidemiology, environmental health sciences, and health policy and management. Her research and professional activities bridge science and public policy, with a focus on the incorporation of epidemiology and mechanistic toxicology into environmental and occupational health policy. Her areas of current focus include the health and environmental impacts of industrial food animal production; cardiovascular risks of arsenic, lead, and cadmium; and immunotoxicity of mercury compounds. Silbergeld has served as a science advisor for the Environmental Protection Agency, Department of Energy, Centers for Disease Control, National Institute of Environmental Health Sciences, Occupational Safety and Health Administration, and international organizations, including the World Bank, United Nations Environment Programme, World Health Organization, Pan American Health Organization, Food and Agriculture Organization, and the International Labour Organization. She has served as editor-in-chief of *Environmental Research* and has published more than 450 peer-reviewed articles, monographs, and reviews. She is the recipient of a lifetime achievement award from the Society of Toxicology, the Barsky Award from the American Public Health Organization, and a MacArthur Foundation "genius" award.

While the problem of unsafe tap water in Flint, Michigan, fueled outrage and better awareness in regard to the hazards of lead in tap water, the problem has existed in city after city for years in the US and in other countries. Our author, a winner of the MacArthur Foundation "genius" grant for her work in identifying preventable causes of human disease related to environmental exposures, points out that problems extend well beyond lead. Many potentially harmful contaminants have yet to be evaluated, much less regulated. Her article examines a number of neurotoxins and related issues as they pertain to brain development.

WE ALL HAVE TO DRINK WATER TO LIVE. But the water that we drink is not always safe, even in countries where we assume it is. Some hazards emit a disturbing odor or discolor drinking water's clarity, while others are not as easily detected. Very often drinking water is unsafe because the sources that deliver it to our homes, schools, and businesses are unprotected. Often, water is unsafe because it contains developmental neurotoxins, or DNTs. These chemicals affect brain development from the prenatal period through childhood, with long-term consequences that affect motor function, learning, and behavior. They not only rob our children of their full potential, but also increase the burdens on society, as stunted neurodevelopment can lead to lower graduation rates, increased crime, and reduced lifetime earnings.[1-5]

These economic costs have been calculated in the billions. Moreover, since cognitive attainment (as typically measured by IQ tests and similar instruments) is a continuous function (Figure 1), the population impacts of DNTs extend over a range of exposures varying in intensity but not in likelihood of occurrence. For this reason, a relatively small shift in mean or median performance can result in major impacts on the extremes of distribution. The study of lead exposures and children's IQ scores cited here, for example, demonstrates that a relatively small reduction of five points in the average IQ may result in a 1.6-fold increase in children with severe impairments (intellectual disability), as well as a decrease of more than 50 percent

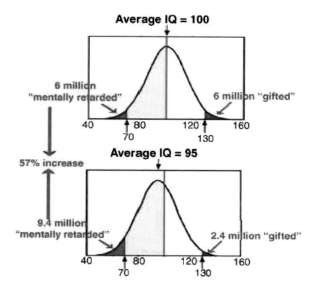

Figure 1. The social implications of developmental neurotoxicity: A slight shift in average IQ scores among a cohort of lead-exposed children results in large increases in children with severe deficits and a loss of children with exceptionally high performance scores. (Source: Fritsch et al. EFSA)

in the number of children with superior intellectual performance (gifted).[6] Both of these have significant impacts on society.

Why is the developing brain so sensitive to neurotoxic hazards, and why do early exposures result in persistent deficits? To start with, humans are distinguished by a highly complex central nervous system that evolved over millions of years. The complexity of our brain requires an extended period of pre- and postnatal molecular events involving a temporal cascade of cell migration, differentiation, and communication, and the biological wiring of neuronal circuitry. Other physiological systems, including the immune and endocrine systems, also contribute to shaping and regulating brain development.

This prolonged process, starting early in gestation and continuing through adolescence, presents an extended period of vulnerability and multiple targets through which harmful exposures can interrupt and alter the developmental sequence required for a normal brain.[7,8] Moreover, since the temporal trajectory of brain development is not fully repeatable, early adverse events can have persistent impacts on brain function as well as delayed effects that may be evident only years later.

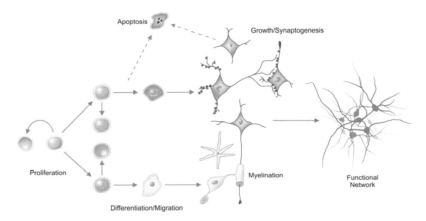

Figure 2. Molecular and cellular targets for developmental neurotoxins over development. (Source: Fritsche et al. EFSA)

Timing and Exposure

The mechanisms by which DNTs affect neurodevelopment and the outcomes associated with these effects are strongly related to the timing of brain development and exposure.[9,10] Generally, DNT exposures early in embryonic development may kill off progenitor cells and reduce cell proliferation, or (in the case of the toxic chemical methylmercury) interfere with cell differentiation and migration. Exposures to DNTs such as pesticides may interfere with apoptosis, or cell death mechanisms essential to the pruning process of neurodevelopment. Later exposures, which have been researched most extensively for certain pesticides and lead, may interfere with intercellular communication and disrupt the establishment of functional networks and synaptic connectivity. The complexity of different timing schedules among brain regions, together with DNTs' multiple modes of action, explain why there can be both agent-specific and age-specific outcomes, resulting in distinctive impairments in cognition, behavior, and motor function.

DNTs also interact with genetics to influence the likelihood and severity of toxicity. This has been clearly demonstrated in recent studies of genetic and acquired risk factors in Autism Spectrum Disorders.[11-13] These environment interactions include polymorphisms, or a discontinuous genetic variation, in genes and epigenetic modifications (external modifications to DNA) of gene expression. Changes in DNA methylation (a "chemical cap"), an epigenetic regulator, have been reported for lead, manganese, arsenic, and mercury.[14] Other DNTs interact with single nucleotide polymorphisms in genes that regulate metal metabolism and distribution in the body. There may also be interactions between certain DNTs and the biome of the gut, which are particularly relevant to drinking water exposure.[15]

DNTs interact with a range of other stressors and exposures that affect brain development. While coincident events and circumstances are important in understanding many outcomes of environmental exposures, this is particularly relevant to neurodevelopmental toxicology. Socioeconomic factors related to health disparities are independently associated with adverse outcomes in cognitive and neurobehavioral development. Many disadvantaged populations, especially children, are also at increased risk of exposure to environmental DNTs, including drinking water contaminants.[16-17] Moreover, children who live in high-risk or underserved communities are often unable to access educational and other interventions that may ameliorate the impacts of early DNT exposures.

Where It Began: Lead

Lead in drinking water was the first environmental agent to be intensively studied for developmental neurotoxicity. It was the Romans who first used lead extensively in vessels for delivering and storing water. While there are no extant medical writings associating toxicity from consumption of water thus contaminated, the Roman architect and engineer Vitruvius cautioned that "water is much more wholesome from earthenware than from lead pipes."[18]

Concerns about lead-containing water systems were raised in early modern times in Germany, Scotland, and the Massachusetts Bay Colony. Writers on medicine and natural history noted that lead and mercury could cause behavioral and neurologic dysfunctions among metal mining and smelting workers. In the 1800s, these metals were already thought to be neurotoxic to both adults and children—that is, to the developed and developing brain—suggesting an early understanding that some mechanisms of toxicity may be age-independent.

In 1977, the first report to clearly associate prenatal lead exposure from drinking water with severe mental retardation came from Glasgow when Sir Abraham Goldberg and his colleagues investigated a cluster of affected young children. Finding no clear cause, they detected prenatal lead exposure by analyzing dried blood spots collected at birth and determined that the source of lead was drinking water contaminated by storage in lead-lined tanks.[19] After achieving regulatory controls on the use of lead in house paints and gasoline, public health investigators are once more assessing the importance of lead in drinking water as a source of medically significant exposures. Recent events in the US have uncovered multiple examples of highly elevated lead levels in drinking water in several cities, notably Flint, Michigan.

DNTs in Drinking Water

Most DNTs in drinking water fall into four broad categories: metals, solvents, industrial chemicals, and natural products. For DNTs such as arsenic and mercury, their presence in water also contributes to food contamination in fish and shellfish and irrigated crops. While most exposures to lead in drinking water—even to the present—result from its use in water delivery systems, other toxic metals, such as arsenic and manganese, get into drinking water from groundwater in contact with geologic formations that contain these elements. This is why arsenic represents a global risk of developmental neurotoxicity in drinking water.[20] Water-borne exposures to most other DNTs are associated with waste disposal and their use in beverage containers. Some of these DNTs, such as mercury, dioxins, and polychlorinated biphenyls, can, like lead, be bioaccumulated in fish and shellfish or taken up from irrigation water by food crops such as rice.[21]

Maternal exposure leads to fetal exposure to many DNTs. It is now well recognized that the human placenta does not protect the fetus from xenobiotic agents that circulate in maternal blood. Some DNTs, including solvents and nanomaterials, can penetrate cell layers;[22] others are actively carried across the maternal-fetal barrier by specific proteins, such as divalent metal transporters that regulate the availability of essential trace elements but can be hijacked by toxic elements.[23,24] The relationship between maternal and fetal exposures is customarily studied by comparing maternal and cord blood. However, these relationships can be more complex for certain DNTs. Lead, for example, may be mobilized from maternal bone stores into the maternal circulation in response to hormonal changes over pregnancy,[25] and chemical compounds such as dioxins mobilized from body fat into maternal blood and breastmilk in association with changes in fat storage.[26]

One of the most potent DNTs is methylmercury, a form of the metal produced in water and soils by bacteria. The links between exposure to this compound and developmental neurotoxicity was first revealed in a catastrophic episode of methylmercury poisoning in the city of Minimata in southern Japan. Minimata more generally demonstrated the importance of prenatal exposures to DNTs and stimulated the growth of the whole field of environmental health. Up to that point, there was little knowledge of mercury effects on children. It took the tenacity of a local physician, Dr. Shoji Kitamura, who fought industrial and government authorities as well as the allegations of other physicians, who concocted the argument that the severe mental retardation and neuromotor deficits in the community were due to genetic deficiencies brought about by intermarriage within the local population.[27] To the contrary, the deficits were due to high levels of mercury in the water.

Two extraordinary observations led him to his diagnosis. The first was his reading of an old textbook on occupational diseases (D. Hunter, *Diseases of Occupation*) that included images of brain tissue from postmortem examinations of workers exposed to mercury. The anatomic damage in these images matched his own examinations of the brains of Minimata children who died early in childhood. His second observation was the dancing cats of Minimata, animals that manifested florid neuromotor and neurobehavioral deficits caused by eating scraps of the same fish consumed by pregnant women in households where infants and children were affected. This observation effectively contradicted claims of genetic insufficiency. His conclusion: Prenatal mercury exposure is associated with distinctive changes in the structure of affected brain regions, particularly the cerebellum and cerebellar cortex. To further explain this, a distinct mechanism for mercury has been proposed by Sass et al., involving immune system regulation of some of the earliest events in neurodevelopment and migration of differentiated cells to these brain regions.[28]

Natural geologic sources are largely responsible for manganese contamination in drinking water. Manganese has a long history of neurotoxicity, first observed in studies of adult workers in manganese mining and smelting. After chronic occupational exposures, many of these workers were diagnosed with a Parkinsonism-like syndrome with both cognitive and neuromotor manifestations. These similarities extend to the efficacy of L-dopa treatment. Later analyses have differentiated manganese intoxication from idiopathic Parkinsonism in terms of the brain regions and neuronal pathways affected.[29]

In children, high levels of manganese are associated with deficits in neuromotor development as well as cognitive impairments. Studies also suggest that there may be some similar adverse outcome pathways for manganese in adults and children.[30] A large study of prenatal manganese exposure—as part of a longitudinal study of environmental risk factors by child development researchers in Korea—found that both high and low manganese concentrations in maternal blood were associated with poorer performance by children tested for early learning at age six.[31] This complex dose-response curve is consistent with the observation that manganese is an essential trace element during early development, but that the range of exposures conferring benefit is narrow.

Arsenic, like manganese, is one of the few metal DNTs for which most exposures begin and largely remain associated with drinking water sourced from groundwater. But like mercury, exposures to arsenic can also occur through consumption of foods, particularly certain rice strains that can bioaccumulate the metal from irrigation water.[20] Arsenic has been less extensively studied epidemiologically for developmental neurotoxicity in children, but the available literature indicates that it is likely to be a DNT metalloid.[32]

Nonmetallic DNTs

Some of the most prevalent contaminants in drinking water are generated by reactions between naturally occurring organic matter in water and disinfection agents containing chlorine or bromine. Health concerns about these byproducts, which include trihalomethanes and haloacetic acids, have focused on potential carcinogenicity. However, an important paper published in 2004 reported that early exposure of rats to dibromoacetic acid caused significant neurotoxicity.[33]

Residues of consumer products—pharmaceuticals, cosmetics, and personal care products—are also widely present in drinking water.[34] These DNTs are discharged in domestic wastewater and reappear in drinking water because current methods of treatment are not adequate to remove them. Among these DNTs are psychoactive drugs and endocrine regulators such as birth control drugs. A recent study of the antiepileptic drug carbamazepine challenges assumptions that the concentrations of pharmaceuticals in drinking water are below the level at which effects on the brain would be expected to occur in the fetus.[35] Concentrations of estrogens in drinking water sources can be sufficient to affect the development of lower animals.[36]

In addition to drugs, other chemicals reach drinking water through household wastes. These include endocrine disrupters, both natural and man-made: a broad range of chemicals that may interfere with the body's endogenous hormone function and produce adverse developmental, reproductive, neurological, and immune effects.[37] Epidemiological and experimental studies have linked developmental neurotoxicity to pre- and early postnatal exposures to several classes of these endocrine disrupting chemicals (EDCs), including phthalates, polyhalogenated organic molecules, and perfluorinated compounds.[37] Among EDCs are chemicals that disrupt neonatal thyroid hormone function, a physiological system essential for normal brain development. Other EDCs, such as those that interact with estrogenic signaling pathways, also have anatomic effects on the brain and genitalia and impact the normal development of sex-typic behaviors in young boys and girls.[40]

Many EDCs have been detected in watersheds in the US and the blood of almost all Americans. Others, such as bisphenol A, leach into drinking water and other beverages from containers, including baby bottles. Phthalates have been measured in beverages and in intravenous fluids delivered by plastic tubing. We have not as yet identified all the EDCs in drinking water; a recent study of surface waters downstream from the discharge of fracking wastes reported multiple interactions with receptors based on assays to detect endocrine activity.[38]

Not all DNTs are products of the chemical industry or the result of human activities. Microscopic organisms in water, including dinoflagellates and algae, can produce diverse and highly potent neurotoxins.[41] We can be exposed through contact and ingestion of water as well as inhalation of bioaerosols carrying these natural neurotoxins. We can also be exposed through the consumption of fish and shellfish that accumulate certain of these neurotoxins, such as domoic acid or saxitoxin.[42] Algal toxins are well characterized as neurotoxic to adults and have been demonstrated to be DNTs in experimental animals and marine mammals.[43] But apart from case reports, there is scant information on health effects in children and adolescents.[44,45]

Ignorance Is Not Acceptable

Because DNTs in drinking water are often undetected, it's no surprise that studies have found that too many children and women of childbearing age in the US and elsewhere are exposed to them. Water delivery systems are complex and can involve lead and disinfectants that produce DNTs. Other types of these compounds, including pesticides, industrial products, and natural toxins, reach drinking water though a variety of environmental releases and pathways. Moreover, DNTs in water can contaminate fish, shellfish, and crops. Natural toxins, such as those produced by phytoplankton, result in highly toxic water, air, and food-borne exposures.

While technology allows us to contemplate voyages to faraway planets and create devices to purchase things with a wave of the phone, we remain largely ignorant of the nature and extent of DNTs in our water, food, and other consumer products. Our knowledge of toxic agents such as methylmercury, arsenic, and lead is largely drawn from devastating effects of population exposures that we haven't prevented.[46] Recent discoveries about endocrine disruptors (perfluorinated compounds (PFCs, for example) illustrate the magnitude of our failure to identify the neurotoxicity of chemicals commonly found in drinking water.[39] Through a survey of workers' blood levels, PFCs were accidentally discovered to be highly persistent and, because of their multiple uses, likely to be almost ubiquitous in human blood samples. Several toxicological studies have reported adverse impacts of these chemicals on brain development, while epidemiological studies indicate associations between prenatal exposures to PFCs and deficits in early childhood development in the US and other countries.[37]

We depend upon toxicological studies to identify potential DNTs among the thousands of existing chemicals in use here and abroad, prior to human exposure. However, our current approaches are widely acknowledged to be inadequate. Testing methods are expensive and often criticized for their lack of specificity and failure to detect the multiple mechanisms involved.[47] That critical outcomes include deficits in learning and social behavior challenges methods utilizing animal models and cells for identification of potential DNTs.

As a result, we have not progressed much from the discovery of lead as a cause of intellectual disability in infants, and mercury as the cause of severe neurotoxicity in young children. In these instances, investigations that identified preventable environmental exposures were only pursued because no other cause was identified. The cases of lead in Glasgow and mercury in Minimata were exceptional, too, because of their severity and the persistent and extraordinary detective work of local physicians. This is not reassuring.

Since then, epidemiological studies of outcomes related to developmental neurotoxicity among children exposed pre- and postnatally have helped us to define the risks of a limited number of DNTs, especially at low-level exposures. The problem is that most of this research continues to focus on a limited set of well-characterized agents and does not adequately lead to studies that will identify the number or variety of DNTs in drinking water and elsewhere.[26] Still, advances are coming from longitudinal studies under way in Europe, Korea, and Japan. Some of this epidemiological research has already contributed critical new knowledge about known toxicants and may reveal additional compounds for further study.[48]

Knowledge of toxic chemicals in our air is much further along than in our drinking water. In the US and many other countries, we have established monitoring systems for airborne hazards and now control many of them at the source of release. We don't do this for drinking water. For a more complete knowledge of exposure to DNTs from this source, we need a national system for environmental monitoring and exposure assessment. Most of the available data on chemicals and other substances in water come from studies by the United States Geological Survey, rather than from the Environmental Protection Agency, and these surveys are not generally designed to assess population exposures. The recent scandal over lead in Flint has revealed the extent to which drinking water needs to be monitored. To prevent developmental neurotoxicity in the years ahead, we must do much better.

8

Making Mental Health a Global Priority

By Patricio V. Marquez, Sc.M., and Shekhar Saxena, M.D.

Patricio V. Marquez, ScM., a lead public health specialist at the World Bank Group (WBG), coordinates the Global Tobacco Control and Mental Health Initiatives. He co-organized the WBG/WHO "Out of the Shadows: Making Mental Health a Global Development Priority" conference on April 13-14, 2016, held at WBG/IMF Spring Meetings. At WBG he has served as public health focal point (2014-15), co-leader of Ebola Emergency Response Program for West Africa (2014-15), and as a member of teams that prepared the Global Avian Influenza Preparedness and Control Program in 2006 and the Global Food Response Facility in 2008. Marquez, originally from Ecuador, has worked in more than 50 countries in Africa, Europe, Central Asia, Latin America, the Caribbean, East Asia, and the Pacific since 1988. He is also part of the Global Work Group of the Advisory Committee to the US CDC director (2015-present). Marquez is a graduate of the Johns Hopkins University School of Public Health.

Shekhar Saxena, M.D., a psychiatrist by training, has worked at the World Health

Organization (WHO) since 1998 and as the director of the Department of Mental Health and Substance Abuse since 2010. He provides advice and technical assistance to ministries of health on the prevention and management of mental, developmental, neurological and substance use disorders, and suicide prevention. His work also involves establishing partnerships with academic centers and civil society organizations and global advocacy for mental health and substance-use issues. Saxena, originally from India, is leading WHO's work to implement the Comprehensive Mental Health Action Plan adopted by the World Health Assembly in May 2013 and scaling up care for priority mental, neurological, and substance-use disorders.

At a conference in April in Washington, D.C., the World Bank Group (WBG), together with the World Health Organization (WHO) and other partners, kick-started a call to action to governments, international partners, health professionals, and others to find solutions to a rising global mental health problem. Our authors write that mental disorders account for 30 percent of the non-fatal disease burden worldwide and 10 percent of overall disease burden, including death and disability, and that the global cost—estimated to be approximately $2.5 trillion in 2010—is expected to rise to $6 trillion by 2030.

WHO HASN'T FELT A SENSE OF LOSS OR DETACHMENT from our families, friends, and regular routines, or experienced nervousness and anxiety about changes in our personal and professional lives? For some, fear and worry constantly distract, confuse, and agitate. For others, frequent and severe bouts of depression are a debilitating daily burden that interferes with family, career, and social responsibilities. All too often, such problems lead to alcohol or drug abuse, self-destructive behavior, or even suicide. Mental health is an essential part of human existence—but it tends to be transitory for millions of people throughout the world.

Mental disorders can also be triggered by massive social dislocations —driven by economic crises, such as the financial crisis of 2008;[1] civil conflicts, war, and violence in places like the Middle East, Central America, and Africa;[2] epidemics like the recent Ebola outbreak in West Africa;[3] or earthquakes, such as the recent one in Nepal.[4] Even after economic growth returns and unemployment drops, peace settlements are made, or we reach zero Ebola cases; after the dead are mourned and the rebuilding of countries gets under way, there is long-term damage left behind in the social fabric of affected communities and the mental well-being of individuals.

The social costs of mental and substance-use disorders, including depression, anxiety, schizophrenia, and drug and alcohol abuse, are enormous.[5] Studies estimate that at least 10 percent of the world's population is affected,[6] including 20 percent of children and adolescents.[7] The World Health

Organization (WHO) estimates that mental disorders account for 30 percent of nonfatal disease burden worldwide and 10 percent of overall disease burden, including death and disability.[8] In addition to their health impact, mental disorders cause a significant economic burden. There is also a notable link between them and costly, chronic medical conditions, including cancer, cardiovascular disease, diabetes, HIV, and obesity.

The global cost of mental disorders was estimated to be approximately $2.5 trillion in 2010; by 2030, that figure is projected to go up by 240 percent, to $6 trillion. In 2010, 54 percent of that burden was borne by low- and middle-income countries; by 2030, the proportion is projected to reach 58 percent.[9] The overwhelming majority (roughly two-thirds) of those costs are indirect ones associated with the loss of productivity and income due to disability or death. Several recent studies in high-income countries have found that the costs associated with mental disorders total between 2.3 and 4.4 percent of gross domestic product (GDP).[10]

It has become increasingly clear that most countries in the world are ill prepared to deal with this often invisible and overlooked health and social burden. In the second decade of the 21st century, not much has changed in how many countries view and deal with mental illness. Some are still using 17th century tactics to "protect society": confining and abandoning the "mad" in asylums or psychiatric hospitals, often for life, which grossly compounds the negative impact on these individuals and on society as a whole.[11] Despite its enormous societal burden, mental disorders continue to be driven into the shadows by stigma, prejudice, and fear that disclosing affliction may mean jobs lost and social standing ruined, or simply because health and social support services are not available or are out of reach for the afflicted and their families.

During April 2016's World Bank Group(WBG)/World Health Organization Global Mental Health Event in Washington, D.C.,[12] hundreds of doctors, aid groups, and government officials convened to start an ambitious effort to move mental health away from the margins of the international development agenda. From the start of the conference, it was evident that, despite enormous challenges inherent in the enterprise, there is growing impatience to move mental health from the periphery to the center of the global health and development agenda. As highlighted in WHO's Mental Health Action Plan 2013-2020,[13] and in the summary report and commentary prepared after the 2016 WBG/WHO event,[14-15] a number of evidence-based interventions have been effective in promoting, protecting, and restoring mental health, well beyond the institutionalization approaches of the past. Properly implemented, these interventions represent "best buys" for any society, with significant returns in terms of health and economic gains. Some of these interventions are within the health sector (e.g., treatment with medicines or psychotherapy) and others outside it (e.g., providing timely humanitarian assistance to refugees).

Economic Loss and Return on Investment

Countries are not investing adequately in mental health; for most, it is not high on their list of priorities.[16] One-third of the countries do not even have a mental health policy or plan and about half do not have a mental health law. Most countries in the low- or middle-income group spend less than $2 per capita on mental health. Many allocate less than 1 percent of their health budget on mental health. The number of trained health professionals delivering mental health care is also grossly insufficient; many countries have less than one psychiatrist for every one-million people. Often, scarce resources are utilized inefficiently. While it is widely accepted that old-style psychiatric hospitals are poorly suited for mental health care, 60 percent of inpatient beds, globally, are still in such institutions.

A study prepared for the WBG/WHO global mental health event[17] (using the estimated prevalence of depression and anxiety in different regions)[18] presents a new projection of treatment costs and outcomes for the 2016-30 period in 36 low-, middle-, and high-income countries that between them account for 80 percent of the global burden of common mental disorders. A modest improvement of 5 percent in the ability to work and in productivity as a result of treatment was factored in and mapped to prevailing rates of labor participation and GDP per worker in each of the 36 countries analyzed. The key outputs of the analysis were year-by-year estimates of the total costs of treatment (the investment), increased healthy life years gained as a result of treatment (health return), enhanced levels of productivity (economic return), and the intrinsic value associated with better health.

The results show that the investment needed to expand effective treatment for common mental disorders is substantial: In the 36 countries for the period in question, it amounts to $141 billion, with $91 billion going toward depression treatment and $50 billion for anxiety disorders. The returns on this investment are also substantial. A 5 percent improvement in labor participation and productivity produces an estimated global return of more than $399 billion, $230 billion of which result from scaled-up depression treatment and $169 billion from better treatment of anxiety disorders. The economic value of improved health is also significant ($250 billion for scaled-up depression treatment alone). The end result is a favorable benefit-to-cost ratio, ranging between 2.3–3.0 to 1 when economic benefits only are considered and 3.3–5.7 to 1 when social returns are also included.

Mental Health Parity in the Global Health Agenda

Moving from theoretical to practical gains would require wider acceptance of the idea that mental health disorders are conditions of the brain that should not be treated differently than other chronic health conditions, such as heart disease or cancer.[19] Nor, in fact, are they truly separable: If untreated, mental disorders can negatively affect management of such co-occurring diseases as tuberculosis and HIV, diabetes, hypertension, cardiovascular disease, and cancer.

In the United States, as well as countries such as Chile, Colombia, and Ghana, attempts to push for mental illnesses and addiction treatment equality come up against clauses that deny health-insurance coverage for preexisting conditions, a common barrier. And when this hurdle is overcome, as explained in a vivid personal account by former US Congressman Patrick Kennedy,[20] the next big issue is determining what is covered and funded at the provider level. And this leads to a host of additional questions, such as what conditions to cover, how to select a menu of evidence-based treatments to be offered by service providers at different levels of care (as is commonly done for other health conditions), and how these services will be funded and reimbursed without perpetuating indirect medical discrimination through high deductibles, copayments, and lifetime limitations in coverage.

This is not an easy task. Strategies and plans for the medium term must be developed across countries to integrate mental health care into health services delivery platforms that focus on the whole patient rather than an aggregation of diseases. And even if these policy and service delivery changes were adopted, the need would remain for unrelenting effort to support affected persons and their families, empowering them to defy the stigma of being seen as "mentally ill" and to get essential services and adhere to prescribed treatments.

Mental Health of Migrants and Refugees

Conflict exposes civilian populations and refugees to violence and high levels of stress,[21] causing dramatic rises in mental illness[22] that can continue for decades after armed conflict has ceased. Cambodians, for example, continue to suffer widespread mental illness and poor health almost four decades after the Khmer Rouge-led genocide of the late 1970s.[23] Rebuilding efforts in postconflict and postdisaster societies, therefore, should include building out mental health services that are well integrated into primary care and public health. A series of catastrophic earthquakes in Japan, including the 1995 Hanshin-Awaji earthquake, the 2006 Niigata Chuetsu earthquake, and the 2011 Great East Japan earthquake, has provided evidence that mental health and psychosocial support can also be effectively integrated into humanitarian response and disaster risk management.[24]

It has been established that in low- and middle-income settings, an emergency provides an opportunity to improve the mental health system.[25] At present, projects funded by the WBG and other organizations utilize a bottom-up, multidisciplinary approach to reintegrate displaced population groups after conflicts and natural disasters. Incorporating treatment for mental illness into such projects would help overcome barriers to employment among the poor and vulnerable. Further investment in education, so-cial protection, and employment training would ameliorate social exclusion and build social resilience by serving the unique needs of vulnerable groups. In Canada, for example, an initiative known as Rise Asset Development is another source of funding and support for persons with mental health problems. Rise provides a combination of low-interest small business loans, training, and mentorship to entrepreneurs with a history of mental health or addiction challenges in order to support their self-employment ambi-tions (and enjoys a 93 percent payback rate).[26]

Technological Solutions

Information and communications technology (ICT) can be a useful instru-ment for global mental health. It offers alternative modes of mental health care delivery when resources are scarce, and new ways to address long-stand-ing obstacles that hinder access to care, such as transportation barriers, stig-ma associated with visiting mental health clinics, clinician shortages, and high costs.[27] These platforms, especially in mobile formats, can offer remote screening, diagnosis, monitoring, and treatment, and remote training for nonspecialist health care workers. They can be instrumental in developing and delivering highly specific, contextualized interventions.[28] Overall, ICT for mental health has a potentially important supporting function for spe-cialized care and community mental health care, and could enhance and enable informal approaches and self-care as well.

Important though digital innovation promises to be, it needs scientific validation before it can become part of global mental health. Data collec-tion achieved through technology is fundamental for advancing evidence in the field. Data collected from individuals will, furthermore, create a basis for strengthening the understanding of mental health and behavioral disor-ders and take that understanding to another level. Timely access to data for decision making can help improve health care organization, allocation of resources, and service delivery.

Governments should work with the private sector, academia, and the medical establishment to develop and adapt these tools to advance the mental health agenda.

Mental Health in the Workplace

There is a robust body of evidence showing that investment in workplace wellness programs is not only good for employees but also for companies' bottom line.[29] In addition to obesity and smoking cessation programs, such interventions commonly focus on stress management, nutrition, alcohol abuse, and blood pressure, and on preventive care such as flu vaccination.[30] In regard to mental health, workplace interventions focused on individuals might center on either treatment or promotion, such as cognitive-behavioral approaches to stress reduction. Organization-level policies can encourage interventions that address prevention and early intervention. There is some evidence that an integrated approach to workplace mental health that includes harm prevention through reducing workplace risks, mental health promotion, and treatment of existing illness provides comprehensive management of mental health needs. A simple guide with seven steps toward a mentally healthy organization has been published by the Global Agenda Council of the World Economic Forum.[31]

Relevance of Neuroscience

At the WBG/WHO global mental health event, such as experts Gustavo Roman, director of Houston Methodist Neurological Institute, emphasized that although mental illnesses are brain diseases, this concept has been lost over the years and ignored by policy makers.[32] To reverse this situation, he advocated calling mental illnesses neuropsychiatric diseases, a term used by WHO, since it helps to address mental and neurological disorders as a group, where mental health is considered along with neurology.

Indeed, advances in research on brain structure and function as well as in molecular genetics have already contributed enormously to our understanding of several mental disorders. For example, certain brain regions and neurotransmitters have been identified as important in depression. Genes that apparently increase the risk of diverse mental disorders have likewise been identified. However, these scientific advances have not yet defined and validated biomarkers that can be used at a population level; they have facilitated development of newer medicines, but not yet resulted in breakthrough discoveries. Several brain projects that have been initiated across the world (e.g., in the US, Europe, Japan, and China) are likely to contribute to more knowledge and better diagnostic and therapeutic tools for mental disorders in the future. But whether these will significantly impact the overall global burden of mental disorders in the near future is not entirely clear.

Collaboration and Financing Options

Jim Y. Kim, president of the WBG, noted during the opening plenary session at the conference that the WBG, together with WHO and other international and national partners, have kick-started an important global conversation and a call to action to governments, international partners, health professionals, and community and humanitarian workers.[33]

The physical, social, and economic burden and cost of mental illness are too large to ignore. Since the impact of mental health is pervasive and relevant to not only health but to other sectors, like education and labor, investing in mental health would significantly contribute to more general efforts to reduce poverty and share prosperity. Indeed, many non-health-related global concerns have clear linkages to mental illness, such as enduring poverty, natural disasters, wars, and refugee crises. Also, such existing health priorities as noncommunicable medical diseases, child health, and HIV are inextricably related to mental health. They provide entry points to link priorities and collaboration with relevant actors in order to increase investment in mental health.

The challenge is clear: If we are to fully embrace and support the progressive realization of universal health coverage, we must work to ensure that prevention, treatment, and care services for mental health disorders at the community level, along with psychosocial support mechanisms, are integrated into service delivery platforms, and are accessible and covered under financial protection arrangements. But we must also advocate for and identify entry points across sectors to address the social and economic factors that contribute to the onset and perpetuation of mental health disorders.

The exploration of alternative sources of financing to support mental health parity in the health system and to mainstream across other "entry points" should be a priority. For example, if development lifts lives, and new and innovative approaches for funding development are seen as "game changers," then perhaps we could argue that the development community, in accordance with the 2015 Financing for Development Addis Ababa Action Agenda,[34] needs to redouble its commitment to advocate with national governments and society at large for raising "sin taxes" such as taxes on tobacco, alcohol, and sugary drinks,[35-38] which are a win-win for public health and domestic revenue mobilization.

For example, taxing tobacco is one of the most cost-effective measures to reduce consumption of products that kill prematurely, make people ill with diverse diseases (e.g., cancer, heart disease, and respiratory illnesses), and burden health systems with enormous costs. In addition, hiking tobacco taxes can help expand a country's tax base to mobilize needed public revenue to fund vital investments and essential public services that benefit the entire population and help build the human capital base of countries, such as financing the progressive realization of universal health coverage, including mental health care. Indeed, data from different countries indicate that the annual tax revenue from excise taxes on tobacco can be substantial: In the US, for example, as part of the 2009 reauthorization of the Children's Health Insurance Program approved by the US Congress, and that President Obama signed as the first bill after being elected, a 62 percent per pack increase in the federal cigarette tax was adopted to help fund the program, increasing the total federal cigarette tax to about $1 a pack. Federal cigarette tax revenue rose by 129 percent, from $6.8 billion to $15.5 billion, in the 12 months after the tax (April 2009 to March 2010), while cigarette pack sales declined by 8.3 percent in 2009—the largest decline since 1932.[39]

In the Philippines, the adoption of the 2012 Sin Tax Law showed that substantial tax increases on tobacco and alcohol is good for public health impact and for resource mobilization for health investments. In the first three years of implementation of the law, $3.9 billion in additional fiscal revenues was collected. The additional fiscal space increased the Department of Health budget threefold and increased the number of families whose health insurance premiums were paid by the national government from 5.2 million primary members in 2012 to 15.3 million in 2015, or about 45 million poor Filipinos (about 50 percent of the total population). Indeed, both country initiatives show that increasing taxes on tobacco and alcohol is a low lying fruit to raise domestic resources to attain sustainable development goals, including expanding mental health care coverage.[40]

As we move forward with this task, we should be guided by the belief that the agonies of mental health problems that blight and distort lives and communities and that impose a heavy economic and social burden on the planet can be dealt with effectively—if there is political commitment, broad social engagement, additional funding, and international support to make mental health an integral part of health care and promotion across the globe.[41]

9

The Human Connectome Project
Progress and Prospects

By David C. Van Essen, Ph.D., and Matthew F. Glasser, Ph.D.

David C. Van Essen, Ph.D., is the Alumni Endowed Professor in the Department of Neuroscience at Washington University in St. Louis. He has served as editor-in-chief of the *Journal of Neuroscience*, founding chair of the Organization for Human Brain Mapping, and president of the Society for Neuroscience. He is a fellow of the AAAS and received the Peter Raven Lifetime Achievement Award from the St. Louis Academy of Science and the Krieg Cortical Discoverer Award from the Cajal Club. Van Essen is internationally known for his research on the structure, function, connectivity, and development of cerebral cortex in humans and nonhuman primates. He led the Human Connectome Project (HCP) and co-leads two Lifespan HCP projects that will acquire and share data on brain circuitry in childhood and in aging.

Matthew F. Glasser, Ph.D., a medical student at Washington University in St. Louis, completed his Ph.D. training with David Van Essen. Glasser has over a decade of experience in brain imaging research with a focus on brain anatomy and brain imaging methods development, and has authored or coauthored 41 peer-reviewed articles. He is best known for his work on reconstructing the arcuate fasciculus, the main connection between the brain's language areas; for developing novel or improved methods for mapping cortical areas, such as mapping the amount of neuronal insulation, called myelin, of the cortical grey matter based on clinical T1-weighted/T2-weighted MRI images; and for producing a new multimodal map of the human cerebral cortex. Glasser is pursuing clinical training in neuroradiology to be a physician-scientist neuroradiologist.

 *As the first phase of one of the most ambitious projects in the history of neuroscience comes to a close, one early and influential leader and his younger colleague explain its evolution and underpinnings. Its goal "is to build a 'network map' that will shed light on the anatomical and functional connectivity within the healthy human brain, as well as to produce a body of data that will facilitate research into brain disorders such as dyslexia, autism, Alzheimer's disease, and schizophrenia."*

UNDERSTANDING THE HUMAN BRAIN in health and disease represents a grand scientific challenge for the 21st century and beyond. How does a collection of 90 billion neurons interconnected by 150 trillion synapses give rise to the extraordinary capabilities of human behavior and the amazing diversity of talents among the billions of people populating our earth? Recent years have seen exciting progress in addressing these fascinating questions, but achieving a deeper understanding of exactly how the human brain functions and what goes awry in various disorders remains a profoundly demanding endeavor. The more we learn, the more we appreciate how much is left to learn.

Progress in neuroscience has benefited greatly from increasingly powerful methods for acquiring, analyzing, and sharing data. For human brain studies, many noninvasive neuroimaging methods have emerged in recent decades. Among these, magnetic resonance imaging (MRI) has become a workhorse technology because of the diversity of information attainable using the same scanner to acquire images. Four main types of MRI are particularly germane here:

- *Structural MRI* provides simple but high-resolution images of the brain, helpful in making geometrical models of brain structures necessary for modern brain imaging analysis and in analyzing subtle aspects of brain architecture, such as the thickness of the cerebral cortex or the amount of neuronal insulation, myelin, within the grey matter.

- *Task-activated functional MRI (tfMRI)* reveals which brain regions are relatively more activated or deactivated during performance of behavioral tasks, based on the blood oxygen level dependent (BOLD) MRI signal, which is modulated by neural activity via a process of neurovascular coupling.

- *Diffusion MRI (dMRI)* analyzes diffusion of water molecules within the white matter to infer the orientation of axonal fiber bundles; this information can then be used to infer long-distance white matter tracts (tractography) that connect distant gray matter locations, often called "structural connectivity."

- *Resting-state fMRI (rfMRI)* correlates spontaneous fluctuations in the BOLD signal while subjects lie in the scanner and let their minds wander, to reveal how parts of the brain activate or deactivate together at rest—thereby enabling estimates of 'functional connectivity.'

Brain imaging studies have typically focused on only one of the above approaches, and it has been challenging to integrate information across approaches and across different studies.

Getting off the Ground

In 2009, leaders at the National Institutes of Health (NIH) recognized an opportunity to build on these advances by promoting the systematic characterization of human brain connectivity and its relationship to behavior. With Michael Huerta (associate director of the National Institute of Mental Health at the time) playing a key role, the NIH Neuroscience Blueprint announced a competition for the Human Connectome Project (HCP), with the overarching objectives of acquiring, analyzing, and freely sharing information about brain circuitry and connectivity gathered by noninvasively imaging healthy young adults. This funding opportunity spurred interest among neuroscientists around the world.

The scope of the project clearly warranted a multi-institutional consortium, leading to a complex courtship among potential dancing partners. A group of us at Washington University (WashU), having strong neurobiological and neuroinformatics expertise, joined with a University of Minnesota (UMinn) team of MR pulse-sequence experts led by Kamil Ugurbil; an Oxford University group of neuroimaging analysis experts led by Steve Smith; and investigators at several other institutions with expertise in magnetoencephalography. In 2010, our "WU-Minn-Ox" consortium received a $30 million HCP award from NIH. A separate HCP grant was awarded to investigators at MGH and UCLA to build a special scanner customized for diffusion imaging.

Six years later, the "young adult HCP" is drawing to a close as successor projects, targeting the full human lifespan as well as diseases of the brain, continue. We aim here to summarize some of the major accomplishments of the HCP, in methods development as well as emerging scientific discoveries, and suggest what the future may hold for human connectomics.

Overall, our goal in connectomics is to map the hundreds of functionally distinct areas, or "parcels" of the human brain and to understand how these areas are connected and how each contributes to our behavior. Additionally, we want to understand how the brain's complex functional systems go awry in neurological and psychiatric diseases. Successfully addressing these challenging questions requires acquiring and analyzing MRI data of the highest quality in normal young adult individuals (the original HCP), developing and aging individuals, and in individuals with these diseases.

What Has the HCP Achieved?

The HCP has made substantial improvements in data acquisition, analysis, and sharing. In aggregate, these advances constitute a new 'HCP-style' neuroimaging paradigm whose seven core tenets[1] offer a modern alternative to the traditional approach that has dominated the field for two decades. Here, we use these tenets as a framework for highlighting various important advances.

Tenet 1. *Using multiple imaging modalities, collect as much data as possible from each subject and from as many subjects as possible.*

To this end, the HCP collected four hours of imaging data per subject: about one hour each for the four modalities of structural MRI, tfMRI, rfMRI, and dMRI in four scan sessions over two days.[2] This is an unprecedented amount of scan time for a large-scale project, but having lots of high-quality multimodal data from each subject yields major benefits, as noted below.

Our original target was to study 1,200 healthy young adults. The final major HCP release, slated for Fall 2016, will meet this target with imaging data from 1,100 subjects, plus out-of-scanner behavior-only data from another 100. Moreover, the HCP subjects include twins and their nontwin siblings, enabling analyses of the heritability of brain and behavioral phenotypes to add invaluable "icing" to this large "cake" of imaging data!

Tenet 2. *Maximize the resolution of imaging data and overall data quality.*

The first two years of the HCP grant were devoted, in part, to improving the scanner hardware and developing better MRI pulse sequences; i.e., the specific sequences of radio frequency pulses and magnetic field modulations or gradients needed to generate an MR image. All subjects were scanned using a customized Siemens 3 Tesla (3T) scanner at WashU that provided stronger gradients, leading to better signal to noise in diffusion imaging. (As an aside, Siemens indicated at the time the scanner would be "one-of-a-kind," but it turned out to be a precursor to the Siemens Prisma, a commercial scanner that also provides high gradient strength.)

Even more important than the physical scanner was the incorporation of "multi-band" (a.k.a. "multi-slice") pulse sequences for fMRI and dMRI.[3-5] In fMRI, the HCP used this to increase spatial resolution as measured by the size of the "voxels" (the equivalent of 2D pixels, but in 3D) to 2 mm voxel size, less than the average thickness of the cortical sheet and down from the conventional 3 mm or higher, as well as the temporal resolution (less than one second, down from more than two seconds), which aids in selective noise reduction (see Tenet 3). For dMRI, it enabled a spatial resolution of 1.25 mm, unprecedented for in vivo human studies where 2 mm had previously been considered high resolution.

Tenet 3. *Minimize distortions, blurring, and noise in each subject's data.*

The unprocessed images coming from MR scanners are substantially distorted, in modality-specific ways. For the HCP, we used state-of-the-art existing methods to remove these distortions, and implemented new or refined methods to achieve further improvements.[4, 6, 7]

Time-dependent fluctuations in the BOLD signal are a mix of (1) the neurobiological signals of interest; (2) unstructured temporal noise arising from random ("Gaussian") variation; and (3) structured noise from subject head movement, physiology (e.g., respiration), and transient artifacts related to MR physics. The traditional paradigm typically uses extensive smoothing in space and time to reduce unstructured temporal noise. However, this comes at a high cost, because smoothing also blurs signals across boundaries—between the gray matter (what we care about in fMRI) and the adjoining white matter and cerebrospinal fluid (CSF), across cortical folds (where voxels may be spatially close but are widely separated from each other along the cortical sheet), and across brain-area boundaries. The HCP-style paradigm keeps smoothing to a minimum, and constrains it within the cortical ribbon (for cortex) or within compartmental boundaries (for subcortical nuclei). Also, unstructured noise reduction can be better achieved by averaging within brain parcels (see Tenet 5).

To reduce structured temporal noise, the fMRI fluctuations are decomposed into "components" using independent components analysis (ICA)[8-10] and then a trained classifier algorithm (called "FIX") automatically distinguishes between signal and noise components, allowing noise-related fMRI fluctuations to be selectively removed. In contrast, the traditional paradigm uses less selective approaches like simply deleting noise-corrupted frames or removing the average MRI signal, which is controversial because this average contains both neural and noise effects.[11,12] Further improvements are an area of active work.[1]

Tenet 4. *Respect the natural geometry of brain structures.*

The traditional neuroimaging paradigm relies heavily on volume-based analyses based on a regular 3D array of voxels. This poses a serious problem for dealing with the highly convoluted, sheet-like cerebral cortex, which is better handled with geometric models that respect its topology. David

Van Essen first appreciated the compelling need for a "surface-based" approach four decades ago when he began studying monkey visual cortex.[13] This initially entailed making cortical "flat maps" (collapsing the 3D cortex onto a 2D map) using a tedious manual pencil-and-tracing-paper method.[14] Two decades later, automated methodology finally emerged, based on computerized cortical surface reconstruction algorithms. For more accurate reconstructions, the HCP used high spatial resolution T1w and T2w scans at 0.7 mm (instead of the traditional T1w-only scans at 1 mm) processed by an adaptation of the FreeSurfer method.[15] We also used the T1w/T2w ratio to generate cortical 'myelin maps' that provide valuable markers of cortical areas in many regions.[7,16,17]

Subcortical nuclei require a different approach because they are "blob-like" rather than sheet-like, and occupy only a small fraction of total brain volume. To compactly represent subcortical gray matter as voxels and cerebral cortex as vertices (thus respecting the natural geometry of each domain), the HCP introduced "grayordinates" as a concept and an associated "CIFTI" file format. CIFTI files offer major advantages in data analysis and visualization.[3] (Parenthetically, we note that cerebellar cortex is a folded sheet, but too thin to segment in individual subjects. Hence, it is, regrettably, treated as voxels rather than surface vertices in the current HCP grayordinate representation.)

Tenet 5. *Accurately align brain areas across subjects and across studies.*

Human brains are remarkably diverse, endowing us with our unique personalities and behaviors. Individual differences are especially pronounced for cerebral cortex as (1) the convolutions vary dramatically across individuals (even in identical twins!), especially in higher cognitive regions; and (2) the boundaries of functionally defined areas vary in relation to cortical folds.

The traditional paradigm addresses individual variability using volume-based registration (alignment) of individuals to an atlas. However, even the widely used high-dimensional nonlinear registration algorithms are far from optimal because they fail to respect the topology of the cortical sheet. Van Essen has long contended that surface-based registration is fundamentally a more sound approach.[18]

The widely used FreeSurfer surface registration method has provided a major advance over conventional volume-based methods in dealing with folding variability.[19] However, FreeSurfer cannot achieve good functional alignment in regions where cortical areas are loosely correlated with specific folds. Two decades ago, Van Essen had mused about using functionally relevant features to improve alignment in such regions.[20] This dream became a reality under HCP auspices, as Emma Robinson and Mark Jenkinson implemented the highly flexible Multimodal Surface Matching algorithm, which we tuned to work well on HCP data, using myelin maps and resting-state networks to align the cerebral cortex across individuals.[3, 21]

Tenet 6. *Use neuroanatomically accurate maps of brain areas.*

A fundamental neuroscientific challenge has been to subdivide the brain into neurobiologically accurate and functionally meaningful areas—i.e., to generate a "parts list." This "parcellation problem" has been one of Van Essen's abiding obsessions over his four decades as a cortical cartographer and quickly became Matthew Glasser's as well after he began using brain imaging to study brain connectivity over a decade ago. For human cerebral cortex, the classical anatomists debated early in the 20th century whether there were about 50 or closer to 200 parcels.[22-24] When Glasser joined Van Essen's lab in 2008, they both were keen to use improved neuroimaging methods to attack the parcellation problem, and achieved their first success with the aforementioned myelin maps.[16] The HCP offered a great opportunity to extend this to a truly multimodal approach using high-quality data. Generating an improved cortical parcellation has been the overarching objective that drove our contributions to the methodological advances described in the preceding five tenets.

Figure 1 shows the fruits of these efforts in the form of the "HCP_MMP1.0" multimodal cortical parcellation.[3] It has 180 areas in each hemisphere, with a striking degree of left-right symmetry. This parcellation was initially generated by a semiautomated approach using data averaged across 210 HCP subjects, including cortical thickness, myelin maps, task-activation patterns, and functional connectivity, plus topographic organization revealed by rfMRI. Importantly, a fully automated areal classifier algorithm enabled highly accurate parcellation of the entire neocortex in individual subjects, even those with atypical topological arrangement of specific areas or who weren't used to make the original map. This method reproduced the original parcellation in an independent group of another 210 subjects.

The HCP_MMP1.0 parcellation provides an invaluable neuroanatomical framework that can serve as a foundation for future brain imaging anal-

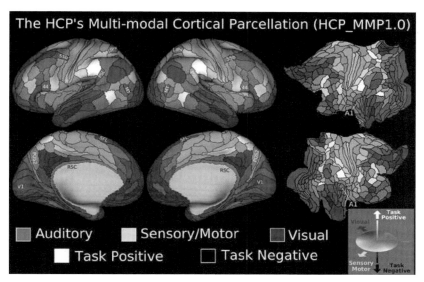

Figure 1. The 180-area per hemisphere HCP_MMP1.0 multimodal human cortical parcellation. With permission, from Ref. 3. Data at http://balsa.wustl.edu/WN56.

yses. It is associated with a treasure trove of multimodal information about each of the 360 areas in the group average and in individual subjects, which is ripe for exploration and data mining using the freely available HCP datasets (see Tenet 7).

Tenet 7. *Freely share data and analysis tools.*

The HCP was mandated by NIH to make its data freely available to the neuroscience community. This was music to Van Essen's ears, as it resonated with his decades-long interest in data sharing, including early efforts with the SumsDB neuroimaging database, which exposed some of the challenges inherent in the enterprise.[25] The HCP provided a great opportunity to take neuroimaging data sharing to a new level. We had the good fortune of teaming up with Dan Marcus (Van Essen's former neurophysiology graduate student), who spearheaded the HCP neuroinformatics effort.

The main HCP database, ConnectomeDB, has served as a highly effective user-friendly platform for sharing the unprocessed and the much more complex (and useful) processed datasets.[26-28] To date, more than 6,000 investigators have signed HCP Data Use Terms and downloaded more than 10 PB (10,000 TB) of HCP data.

There is also a growing need to share extensively analyzed neuroimaging data, particularly datasets associated with published figures—something ConnectomeDB was not designed to handle. Our lab recently launched the "BALSA" database (http://balsa.wustl.edu), which builds on what worked well in SumsDB. BALSA is designed around "scenes" that store all of the information needed to recapitulate exactly what is seen in a complex image viewed on a visualization platform.[29] For example, the data used to generate Figure 1 was uploaded into BALSA exactly as shown in the figure (including all of the annotations); the URL in the figure legend links directly to the corresponding scene in BALSA, and an archive containing 217 MB of associated data can be immediately downloaded. Currently, Connectome Workbench is the only neuroimaging platform that supports scene files and is compatible with BALSA, but this will hopefully change as other software developers recognize the utility of the scene-based approach to data sharing.

The number of publications acknowledging the use of HCP data currently exceeds 140 and is rapidly growing. A few brief examples illustrate the scope and diversity of emerging discoveries: (1) A comparison between resting-state functional connectivity and numerous behavioral and demographic measures revealed a "positive–negative" phenotypic axis that covaries with the strength of functional connectivity in brain regions implicated in higher cognitive functions.[30] (2) An analysis of resting-state functional networks revealed that regions of association cortex tend to overlap with at least two networks, whereas somatosensory and visual regions are more isolated.[31] (3) A model that links resting-state fMRI measurements to task activations in one set of subjects can be used to predict task activation maps for other subjects, suggesting that brain connectivity and function are coupled at the level of individual subjects.[32] (4) Functional connectivity revealed in HCP data correlates with gene expression patterns in postmortem human cortex.[33]

The Present and Future of Human Connectomics

A number of large-scale projects are now building upon the success of the HCP by using its paradigm to study the brain in health and disease. One set, Lifespan Human Connectome projects, focuses on the healthy brain across the full human lifespan. It includes the Lifespan Development project, which will study some 1,300 individuals in ages ranging from five to 21, and the Lifespan Aging project study, which will study 1,200 individuals older than 36.

Each project will include longitudinal scans on a subset of the population, complementing the cross-sectional approach and increasing the sensitivity for characterizing age-related changes. Earlier stages of development are being covered by the Lifespan Baby Connectome project, which will enroll 300 children (from birth to age five) and by the Developing Human Connectome Project in the United Kingdom, which is studying prenatal and neonatal brain development.

Complementing these Lifespan Connectome initiatives is a diverse set of projects focusing on brain disorders, under the general umbrella of the Connectomes Related to Human Disease effort. More than ten have been funded to date, and more are anticipated. They are using the HCP-style paradigm to address many diseases and disorders, including schizophrenia and Alzheimer's disease.

Data sharing for the above projects will be under the umbrella of the Connectome Coordination Facility (CCF; http://humanconnectome.org/ccf). As a major extension of ConnectomeDB, the CCF will provide a common gateway for investigators wanting access to these datasets. Other large-scale projects, such as the recently launched ABCD (Adolescent Brain Cognitive Development) project, will also emulate the HCP-style paradigm, but will use a different data sharing venue.

Importantly, all of these projects entail significant adaptations in data acquisition to deal with practical limitations. Because the original HCP's "luxury" of obtaining four hours of imaging data over a two-day visit is in general not feasible for these other projects, they aim to acquire about 90 minutes of imaging data per subject.

Studying younger and older populations and populations with brain disorders also entails using shorter individual scan durations than the 15 minutes tolerated by the healthy young adults of HCP. We can be optimistic that these "HCP-fast" acquisition protocols will be sufficient to increase the sensitivity for characterizing what may be subtle changes across the lifespan and differences between disease and control groups. However, it will remain critical to be vigilant in distinguishing neurobiologically interesting phenomena from artifacts related to such potential confounds as group differences in head motion or physiological noise inside the scanner. Efforts to better compensate for such confounds and further improve MRI data quality will be important to maintain and accelerate progress.[34]

Finally, and on a personal level, the opportunity to contribute to the HCP and to several successor projects represents a true highlight of Van Essen's decades as a neuroscientist and a cortical cartographer; it provided Glasser with a unique opportunity to learn from an international consortium of experts while making significant contributions to the effort. In our opinion, the HCP represents team science at its best. Its success reflects an exceptionally talented and dedicated group of well over 100 investigators and technical staff who energetically debated ideas and options, then enthusiastically pulled together to carry out the hard work of gathering, analyzing, and sharing a rich dataset that should serve the neuroscience community well for many years to come.

Acknowledgments. We thank Sandra Curtiss for comments on the manuscript. Supported by the Human Connectome Project, WU-Minn Consortium (1U54MH091657) funded by the 16 NIH Institutes and Centers that support the NIH Blueprint for Neuroscience Research; NIH F30 MH097312, MH-60974, and the McDonnell Center for Systems Neuroscience at Washington University.

10

The Evolving View of Astrocytes

By Philip G. Haydon, Ph.D.

Philip G. Haydon, Ph.D., is the Annetta and Gustav Grisard Professor and chair of the neuroscience department at Tufts University School of Medicine. He received his Ph.D. from the University of Leeds, UK. While a faculty member at Iowa State University in 1994, he performed a landmark study by demonstrating that astrocytes release chemical transmitters. In 2001 he joined the faculty of the University of Pennsylvania School of Medicine and, in 2008, moved to Tufts. For 25 years his research has addressed the roles played by glial cells. His recent focus is on the use of glial targets as therapeutic interventions for brain disorders. Haydon has received the Alfred P. Sloan Scholarship, the McKnight Innovator Award, and the Jacob Javits Award from the National Institute of Neurological Disorders and Stroke.

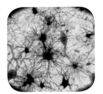 *Did you know that glial cells are more numerous than neurons in the brain? Scientists have found that one type of glial cell that is prevalent in the cortex—the astrocyte—communicates with its brethren, sends information to neurons, and controls blood flow to regions of brain activity. Because of all these properties, and since the cortex is believed responsible for cognition, the role of astrocytes in sleep, learning, and memory is being determined.*

FOR MUCH OF THE 20TH CENTURY, our ability to decode brain signals has been limited to recording and stimulating the electrical activity of neurons. But in the 1990s, more powerful microscopes allowed neuroscientists to observe dynamic, second-to-second chemical changes in the brain. As a consequence, they observed alterations in the levels of chemical signals within certain nonneuronal cells—astrocytes—that fueled a revolution in the understanding of brain function.[1]

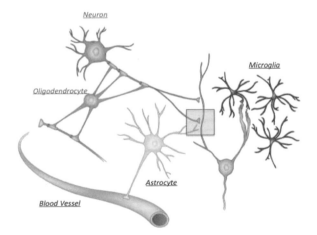

Figure 1. In addition to the electrically active neurons, the brain contains numerous glial cell types, including astrocytes, oligodendrocytes, and microglia. Astrocytes are the most plentiful of these glial cells and have unique physical attributes: They contract blood vessels as well as neuronal synapses at a structure called the tripartite synapse (rectangular box and see Figure 2). Consequently, they play important roles in synaptic development and modulation/homeostasis as well as in the delivery of nutrients from the circulation to neurons and act as an intermediate to relay neuronal activity to the vasculature to control blood flow.

In addition to the electrically active neurons, the brain contains several nonneuronal glial cells (glia is Greek for "glue") as depicted in Figure 1. Astrocytes are the most numerous glial cell type and comprise over half of our brain volume. Although they were once thought to be merely the material that supports brain structure, like the mortar that holds a house together, we now know that they are quite active and intimately involved in signal transmission in the nervous system, responding to and instructing neuronal activity.[2] Because of work performed in the past 25 years, we know that by modulating neural excitation, inhibition, and synaptic transmission, astrocytes influence sleep, learning, and memory,[3-6] and that dysfunction in these cells can lead to debilitating disorders.

Astrocytes' Role in Brain Function

It was soon realized that their physical linkages to both the vasculature and synapses give astrocytes the potential for significant roles in brain function,

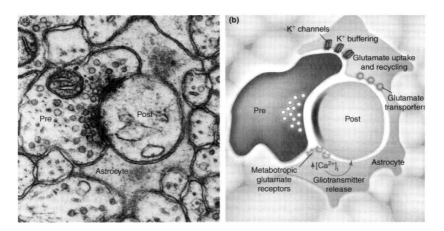

Figure 2. The astrocytic process is the third active element forming the tripartite synapse. (a) Electron micrograph showing a presynaptic (Pre) and postsynaptic (Post) terminal enwrapped by the astrocytic process (green) forming the tripartite synapse. (b) The close association of the astrocytic process with the presynaptic and postsynaptic terminals exerts crucial roles in clearing K+ ions that accumulate following neuronal activity, and in the uptake of the synaptic transmitter glutamate by the activity of plasma-membrane glutamate transporters. Additionally, neurotransmitter release from presynaptic terminals can activate astrocytic receptors that induce Ca^{2+} elevations, which in turn triggers the release of gliotransmitters from these cells.[7] (Courtesy of Cell Press)

such as regulating blood vessel dilatation and modulating synaptic function at a structure known as the tripartite synapse (Figure 2).[3]

With the development of methods that allowed intracellular imaging, several research groups observed that astrocytes exhibited calcium encoding, in which rapid shifts in calcium ion concentration transmit information from one part of the cell to another; they noted how application of the synaptic transmitter glutamate induces calcium encoding signals in astrocytes[2] that stimulate such processes as the release of neurotransmitters from nerve terminals, muscle contraction, and hormone secretion. Changes in calcium ion levels in astrocytes thus gave scientists their first indication that glial cells were active participants in brain function, paving the way to new insights.

Researchers reached another major milestone when they determined that astrocytic calcium encoding signals could induce neuronal activation mediated by glutamate receptors.[8] Within a few short years, we had advanced from knowing little about astrocytes to identifying bidirectional signaling between neurons and these nonneuronal cells, at least in cell culture.[9-11] Subsequent studies examined freshly isolated slices of living brain tissue,[12-14] but the relative imprecision of these techniques limited researchers' ability to draw conclusions about the operation of these signaling pathways. If new insights about the role of astrocytes in neuronal networks were to emerge, new methods needed to be developed.

Modulating Nerve Terminals

The first of these came in 2005, when a cell-specific molecular genetic method enabled researchers to probe the functional role of the astrocyte within the context of neural networks.[15] In this study, a promoter (the DNA region that initiates the synthesis of gene products) that is turned on selectively within astrocytes was used to provoke the expression of proteins that inhibit the release of chemical transmitters from astrocytes. With this approach, my laboratory was able to show that the astrocyte continuously modulates nerve terminals through a chemical signal called adenosine.

Adenosine is known to promote drowsiness (we drink beverages containing caffeine, an antagonist of adenosine receptors, to feel more awake).

Hence we used a mouse model, genetically modified to inhibit the release of adenosine from astrocytes, to determine whether sleep homeostasis (the balance between sleep and wakefulness) was consequently perturbed. As expected, the drive to fall asleep was impaired when we thus prevented adenosine-mediated gliotransmission,[16] an effect that is replicated by pharmacological inhibition of adenosine receptors in normal mice. Over a 15-year time span, we were able to make the transition from identifying bidirectional signaling between astrocytes and neurons to demonstrating roles for these glia cells in the modulation of synaptic activity and a specific behavior. In support of this exciting observation, others have replicated the contribution of astrocytes to sleep by stimulating the astrocyte by optogenetics, rather than inhibiting it. When activated by light, astrocytes provoked slow oscillations in neurons,[17] a result that is compatible with the consequences of genetic inhibition of gliotransmission cited above. Together, these studies have led to the predominant view that astrocytes are generally integrated within neural circuits, where they modify brain function and influence behavior.

Synapse Development

One of the many mysteries of the brain concerns the cues that guide and regulate its development. Of particular interest is this question: Which signals determine whether two neurons are able to functionally connect with one another at a synapse? Some early clues to the answer came from cell-culture studies where scientists plate young neurons into a plastic culture dish and observe whether synaptic connections are able to form. It was found that in the absence of astrocytes, synapse formation was hindered, while in mixed cultures containing both neurons and astrocytes, exuberant synapses formed between neurons. This led to the concept that astrocytes regulate synapse formation. Elegant studies have since shown that proteins called thrombospondins (TSPs) are released from astrocytes and regulate specific aspects of synapse development. The addition of pure TSPs to neuronal cultures obviates the need for other astrocyte factors, and depletion of TSP from astrocytes impairs the ability of this medium to stimulate synapse formation.[18,19]

Interestingly, gabapentin, a therapeutic agent used in the treatment of neuropathic pain and epilepsy, inhibits the receptor for TSPs, suggesting that it may exert its action by preventing new synapse formation in these pathological conditions. In support of this idea is the observation that injury to the cortex, which normally leads to the development of hyperexcitability, is attenuated by treatment with gabapentin.[20]

Astrocytic Metabolic Support

The primary source of energy in the brain is glucose. Astrocytes project a process (called an end foot) that wraps around the blood vessels in the brain to take up glucose from the circulation. But neurons do not make this kind of physical contacts with the vasculature; how then do they obtain energy? One prominent theory holds that astrocytes metabolize the glucose they take up from the circulation to lactate, then transfer it to neighboring neurons that use it as an energy source. This lactate shuttle hypothesis[21] has generated significant debate, but several intriguing observations support it. In particular, astrocytes and neurons express transporters (proteins in a cell membrane that bind to a molecule and transfer it across the membrane) that promote the transfer of lactate from astrocyte to neuron. Additionally, in glucose-depleted conditions, lactate has been shown capable of supporting the energy demands of the neuron.

Cerebral Blood Flow Control

As discussed above, astrocytes, not neurons, predominantly contact the vasculature. Yet neuronal electrical activity induces changes in blood flow. How is this achieved in the absence of direct interactions with the vasculature? Since astrocytes make physical contact with both neurons and the vasculature, they are physically positioned to act as a signal conduit from the first to the second. Since neuronal activity can induce calcium changes in astrocytes, scientists asked whether direct stimulations of calcium changes in astrocytes would lead to changes in blood vessels. To answer this question,[22] initial studies were performed in isolated slices of brain tissue and an optical technique was used to elevate calcium. A chemically synthesized calcium "cage" was introduced into the astrocyte. When light was turned on, the cage opened, liberating calcium in the astrocyte to stimulate calcium-dependent biological events. When this optical technique was used to stimulate calcium dependent events in the astrocyte, the adjacent blood vessels dilated.

Mechanistically, it has been demonstrated that this neuron-astrocyte-vasculature interaction derives from calcium-dependent stimulation of an enzyme called phospholipase A2 (PLA2), which leads to the production of arachidonic acid and its downstream signaling cascades. After initial pharmacological studies supported this idea, when the gene for PLA2 was deleted, astrocytic calcium signals no longer evoked changes in vessel diameter.

Alexander's Disease

Alexander's disease is often referred to as a disease of the astrocyte because it is known to follow from mutations in an astrocytic protein, glial fibrillary acidic protein (GFAP).[23] Alexander's disease is lethal, generally during childhood. Mutations in the GFAP gene are correlated with more than 90 percent of the cases of this disorder, and nearly a third of these cases result from mutations at one of two amino acid sites within the GFAP protein.

A consequence of these mutations is that GFAP, which normally forms fine filaments, instead causes the formation of thick bundles of what are called Rosenthal fibers. The current idea is that mutations in the GFAP protein endow it with new functions such as, according to one prevailing view, binding or sequestering proteins that normally protect the cell. Once sequestered, these proteins cease to be protective, and cellular damage and toxicity result. With these insights, researchers are trying to determine whether they can reduce the amount of the GFAP protein and thus reduce the sequestration of the protective proteins, or alternatively increase the amount of these proteins so they can regain their normal role.

Glutamate Homeostasis and Neuroprotection

A critical function of the astrocyte is helping control the level of extracellular neurotransmitters. The excitatory transmitter glutamate plays pivotal roles in signal transmission in the brain, but its accumulation in the extracellular space can lead to neurodegeneration/toxicity mediated by the persistent activity of the N-methyl D-asparate (NMDA) receptor. These glutamate receptors allow calcium influx into neurons, an important stimulus for learning and memory. But their persistent activity leads to continuous accumulation of calcium, with neurotoxic consequences. Indeed, widespread neuronal death following stroke is, in part, due to this process during and following injury.

The clearance of glutamate from the synapse was one of the first prominent functional roles identified for astrocytes: they abundantly express glutamate transporters, particularly GLT-1, which removes the neurotransmitter from the extracellular space. Deleting this transporter from astrocytes leads to epilepsy and reduced lifespan, reflecting the importance of astrocyte-mediated glutamate control.[24]

Glutamine and Epilepsy

Once glutamate is taken up into astrocytes by glutamate transporters, it is converted to glutamine by an enzyme called glutamine synthetase (GS). Glutamine is an important precursor of both glutamate and the inhibitory neurotransmitter GABA. In certain types of epilepsy, the amount of GS in the brain declines significantly.[25] The consequences of GS loss for brain function were elucidated when, through a technical accident, researchers found that the amount of GS was reduced in normal mice when a specific virus infected astrocytes. This accidental observation provided an opportunity to evaluate the consequence of reduced astrocytic GS on brain function.

Through recordings of synaptic connections between neurons, researchers found an impairment of GABA-mediated inhibitory synaptic transmission when astrocytes did not synthesize GS, which disrupted the balance between excitation and inhibition. An analogy might be cars driving in a city with brake pedals [inhibition] disabled and gas pedals [excitation] intact. The brain circuits involved became hyperexcitable and discharged epileptic-like electrical events. Importantly, the addition of glutamine, which is normally synthesized by GS, restored the normal level of inhibition and effectively put a brake on the mass excitation and epileptic-like discharges.[26] Thus it appears possible that a primary deficit in astrocytic GS leads to a loss of inhibition that results in epileptic seizures.

Amyotrophic Lateral Sclerosis (ALS)

ALS is a progressive disorder in which the motor neurons that cause muscle contractions degenerate, ultimately leading to paralysis and an inability to breathe. The disorder references the baseball player Lou Gehrig, who suffered from ALS. There are two forms of the disease—familial (5-10 percent of cases) and sporadic (90-95 percent). The familial form is caused by mutations in the genome and thus can be inherited from affected individuals. In fact, the mutations in the genetic code for certain proteins causing ALS have been identified, allowing researchers to replicate miscoded genes in mice to further our understanding of the disorder.

For many years, the primary cause of ALS was thought to be within the neuron that degenerates. Cell-culture studies have allowed this idea to be tested since it is possible to introduce neurons and their glial partners, alone or in combination, into the culture dish. When neurons from normal mice were plated into cultures together with astrocytes from mice that carried a familial mutation, the neuron was found to degenerate. Thus, the genetic mutation causes changes to the astrocyte that promote the demise of the neuron, potentially by modulating inflammatory pathways. Further work will allow us to identify potential ways to treat such disorders of the nervous system.

APOE4 and Alzheimer's Disease (AD)

There are two general forms of Alzheimer's disease—early onset (1-5 percent of cases) and late-onset (95-99 percent of cases). Apolipoprotein 4 (APOE4) is the largest known genetic risk factor for late onset AD. Apolipoproteins are expressed by astrocytes and can come in three different forms, in which slight changes in the genetic code have led to small changes in amino acid sequences. These different apolipoproteins are called APOE2, 3, or 4. APOE2 is protective in regard to AD; APOE3 is a mild risk factor for late-onset AD, and APOE4 increases AD risk substantially. For example, individuals carrying two copies of APOE4 are 15 times more likely to get late-onset AD than APOE3 carriers; subjects with two copies of the APOE2 gene are 40 percent less likely to develop the disease than APOE3 carriers. About 40 percent of patients with late-onset AD carry the APOE4 gene.

Mice have been developed in which human APOE genes replaced the mouse APOE genes. This has allowed investigation into the mechanism by which these mutations change the risk of developing AD. One study showed that APOE2 increases while APOE4 reduces the ability of the astrocyte to clear cellular debris, leading the authors to speculate that reduction in this function underlies APOE4-linked AD risk via an increase in synaptic vulnerability and neurodegeneration.[27] Because the APOE4 allelic variant is the greatest known risk factor for late-onset AD and, because astrocytes express the product of this gene, there is intense scientific focus on how astrocytes and APOE4 contribute to this form of dementia.

Over the past 25 years, our understanding of the biology of astrocytes has increased tremendously. Where we initially thought astrocytes to be simple static cells that support neurons, we now know that they listen to and detect the activity of neurons, signal the vasculature to increase local blood flow to provide new nutrients as needed, and act as a conduit to provide energy-rich chemicals to active neurons. We've learned that they also provide feedback signals and regulate processes such as sleep. We are realizing that defects in astrocytes may be the primary signal that causes neuronal degeneration in such diverse disorders as ALS, epilepsy, and Alzheimer's disease. In the near future, it would not be surprising to see the development of new drugs for neurological and psychiatric disorders that target known astrocytic pathways, to protect neurons from neurodegeneration and, ultimately, increase quality of life.

11

Understanding the Terrorist Mind

By Emile Bruneau, Ph.D.

Emile Bruneau, Ph.D., is a researcher and lecturer at the Annenberg School for Communication at the University of Pennsylvania. Prior to his formal training in neuroscience, Bruneau worked, traveled, and lived in a number of conflict regions: South Africa during the transition from apartheid to democracy, Sri Lanka during one of the largest Tamil Tiger strikes in that nation's history, Ireland during "The Troubles," and Israel/Palestine around the Second Intifada. Bruneau is now working to bring the tools of science to bear on the problem of intergroup conflict by characterizing the (often unconscious) cognitive biases that drive conflict, and critically evaluating efforts aimed at transcending these biases. In 2015, he received a Bok Center Award for teaching at Harvard and was honored with the Ed Cairns Early Career Award in Peace Psychology. His work has received funding from the UN, US Institute for Peace, Soros Foundation, DARPA, ONR, and DRAPER Laboratories. Bruneau received his doctorate from the University of Michigan.

*While early research focused on the political roots of terrorism, many of today's investigators are probing the psychological factors that drive adherents to commit their deadly deeds. Are terrorists mentally ill or do they rationally weigh the costs and benefits of their actions and conclude that terrorism is profitable? Our author traces recent advances in using imaging and experimental research to determine what motivates monstrous acts.*

IN THE PAST 15 YEARS, DRAMATIC ACTS OF TERROR have been committed against citizens of many countries. A reasonable first step toward addressing such violence is understanding where it comes from—what motivates people to join terror organizations and engage in terrorism. Recent work in experimental psychology and cognitive neuroscience provides some perspective on the mind of a terrorist.

First, it is useful to define what we mean by "terrorism." When Americans and Europeans think of terrorists, they likely imagine Muslim extremists. For example, Google searches for the term "terrorist" in the month after the Boston Marathon bombings and Paris attacks (committed by Muslim extremists) increased threefold and sixfold, respectively, relative to the months prior to the attacks. By contrast, similarly deadly attacks by Anders Breivik in Norway, who killed 69 children, and Dylann Roof, who killed nine black parishioners in the American South, were not followed by an increase in such searches. In fact, compared to the three months leading up to it, there was a slight decrease in searches using the term "terrorist" after the Breivik attack.

Although it may be comforting to think of terrorists as people unlike us, I will argue that this belief belies an uncomfortable reality: the psychological processes that drive an individual to engage in terrorism are deeply human, common across cultures—and traits that likely reside in us all.

The definition of terrorism that I use here will include two key elements. First, it involves a group ideology. Individuals may attack, threaten, terrorize, or kill others, but if they are not part of a group and not motivated to do so by an ideology, then by this definition they are not terrorists. Even a violent group, such as a drug cartel that beheads civilians, would not be considered a terrorist organization, since its members are not ideologically motivated (note, however, that many of the processes that I describe below also apply to gangs). Second, terrorism is defined by the use of violence in the service of the group's ideology, and particularly violence that indiscriminately targets members of a group (e.g., civilians, children). Many governments would challenge this point, as they have convicted nonviolent Native and environmental activists who sabotage logging equipment under anti-terrorism laws. However, I do not consider these groups to be composed of "terrorists" (but again, many of the processes described below still apply).

In the context of this definition, I will try to offer some insight into the mind of a terrorist by looking at what lies in the human mind more generally. Specifically, I will seek to explain why and how individuals support or engage in "indiscriminate violence driven by group ideologies" by looking at our understanding of three processes: (1) how our brains respond to groups, (2) how our brains are led to condone or initiate acts of indiscriminate violence, and (3) how our brains process ideological information.

Social Factors

For most of the millions of years that our species has been around, humans have eked out an existence only through the coordinated effort of small, cohesive coalitions. Evolution has therefore shaped within us a deep desire to belong to groups. In modern times, social belonging remains a major psychological need, which we fill by connecting with others through a variety of "social identities"—Californian, professor, rugby player, progressive, vegetarian, for example. We each contain multitudes. The tendency to connect through one of these multitudes can be reflexive ("Oh wow, you're from California, too?").

From this perspective, the appeal of "terrorist groups" is completely unremarkable. Just as a fraternity, team, club, military unit, or gang can provide a deep social connection with others, so too can ISIS, Al-Qaeda, or white nationalist groups. Many think that people join groups for what they do (terrorists join terror groups because they are violent people; men join fraternities because they drink and party), but the deep, fundamental motivation to join any group is the need to socially connect. From this view, individuals most at risk for joining a terrorist group are not those who are poor or violent, but those who are alienated and thus drawn to an arrangement that can offer the camaraderie, brotherhood, and purpose that they are missing. This may help explain why very different demographics—the young man at a refugee camp who is deprived of regional, professional, and academic identities, and the middle-class child of immigrants in a Western country who feels alienated from his or her host country—are common recruits for ISIS. And perhaps other groups. It may also explain why regular attendance at a mosque—which provides a strong social identity—is inversely correlated with ISIS enrollment.

Although social identities can in themselves provide clear paths to bring individuals together, the brain seems particularly prone not only to creating an "us," but also readily defining a "them." A classic study that demonstrated the ease with which group identities arise came from a team of experimental psychologists in the 1950s, led by Muzafer Sherif. In the study, the researchers aimed to generate and dissipate group conflict in a set of middle-class white boys who attended a camp set up by the researchers at a park in Oklahoma. The plan was to separate the boys into two groups and then organize a series of activities to establish competitive group identities. In fact, the participants preempted the researchers' strategy: The boys caught wind of each other and immediately formed their own group identities (the "Eagles" and the "Rattlers") and started competing on their own—staking territory, raiding cabins, and picking fights. The silver lining of the study came when the researchers demonstrated how readily they could undo the group distinctions that they had facilitated. By orchestrating a series of threats to the entire camp (a "broken" well, a stuck van) that could only be solved by working together, the stark distinctions between Eagles and Rattlers began to fade as they all adopted the overarching identity of "campers."

Inspired in part by the Sherif study, a host of experimental studies have demonstrated the ease with which people start thinking in terms of "us" and "them," and the consequences of such distinctions.[1] For example, this research has found in controlled lab environments that people assigned to groups based on arbitrary distinctions (e.g., whether they are "underestimators" or "overestimators" of the number of dots on a screen) perceive members of their group to be more intelligent, trustworthy, and attractive than those from the other group. Even if people are explicitly told that the groups are arbitrarily assigned, their minds lead them to assign in-group members higher value than out-group members.

The Impact of Imaging

With the advent of neuroimaging, we have begun to access the inner workings of some of these group-based processes. This has been particularly important, since the mental events associated with "us" versus "them" thinking are likely unconscious, and therefore difficult to assess through self-report. For example, a number of research groups have demonstrated an "out-group race face" bias in a brain: When white participants see pictures of black versus white faces, they register more activity in the amygdala, a brain region that drives fear learning.[2] Since the amount of neural bias is unassociated with explicit anti-black attitudes, this has been taken as evidence that when confronted with a black American, white Americans experience an automatic fear response that they are unaware of and do not necessarily condone.[3]

A particularly interesting iteration of this in-group/out-group face bias illustrates the malleability of in-group/out-group distinctions: Jay van Bavel and colleagues showed that the bias in amygdala activity among white participants was also present more toward an arbitrarily defined outgroup versus the arbitrarily defined in-group, even if the groups were mixed-ace.[4] That is, among white Americans, there was more activity in the amygdala when viewing black versus white faces, but when the same faces were assigned to mixed-race teams, the amygdala now responded more strongly to faces from the "out-group" than it did to faces from the "in-group," regardless of race. We not only have a tendency to generate "us" and "them," but who qualifies for each can be completely flexible, and race/skin color is just one of many arbitrary dimensions over which people can be categorized.

A study published just recently examined the neural basis of another psychological process that has been shown to be distributed parochially: trust. Here, Zaki and colleagues had participants play trust-based economic investment games with in-group and out-group members—in this case, own or rival school members.[5] They found that trusting in-group members resulted in more activity in brain regions associated with pleasure, while trusting out-group members resulted in more activity in brain regions normally associated with cognitive effort (e.g., consciously withholding a response that you desperately want to give, or reassessing a situation). The implication here is that in-group trust comes easy, while out-group trust comes only with effort.

Together, such psychology and imaging studies give us some insight into our genetic legacy. We have inherited brains that are inherently sensitive to group affiliation. We find meaning in our lives through social identities, and we experience comfort with those who share these identities. However, when creating an "us," the brain seems to seek out a "them," bringing online a series of psychological processes—including fear and distrust—that colors our view of out-group members.

Although this schematic helps to illuminate some of the underlying dynamics that may drive people to join a terror group ISIS and groups like it are not merely fraternities or clubs. Terror groups also have an explicit ideology, and membership carries with it a tacit willingness to kill civilians. What neural and psychological processes help us understand the willingness to attach to ideologies, and to condone violence?

The Brain on Ideology

One of the most striking characteristics of terrorist groups is their strict adherence to an ideology. Ideologies provide a narrative structure with which to interpret new information and past events. Since terror groups (without exception, I believe) are composed of an aggrieved minority, their ideology is often centered around a narrative of victimhood.

Such narratives seem particularly powerful; especially, perhaps, for parochial altruists—people who love their own group so much that they are willing to die on its behalf. If you perceive that your group's back is against the wall, this might be just the thing to motivate a parochial altruist to act on its behalf.[6] Perhaps this is why we see the narrative of victimhood even among some of the most powerful groups in the world. For example, note that the "don't tread on me" American flag is still widely visible in the US. In fact, groups often compete with each other for who is the aggrieved victim in a conflict (i.e., "competitive victimhood"), which buys them more third-party support, but also may motivate their members to action.[7]

Whether about victimhood or not, ideologies are incredibly persistent. Part of what gives them their momentum is a set of cognitive filters that helps process incoming information to support and enhance the in-group's ideological narrative. For example, confirmation bias describes people's tendency to uncritically accept information that confirms their group's beliefs, and scrutinize anything that runs counter to their ideological leaning. Certainly, anyone who has paid any attention to the current US election cycle has seen this at play. Another critical bias concerns the way that we construe the deviant actions of others. If I find myself doing something wrong (e.g., cutting late into a merging lane), it is easy for me to justify this by external circumstances (e.g., "I was late for an important meeting"). But when I see others doing the same, I tend to attribute this to their internal characteristics (e.g., "they are selfish jerks"). As was famously expressed by the comedian George Carlin, "Have you ever noticed that anybody driving slower than you is a moron, and everybody driving faster than you is a maniac?" The intergroup context only magnifies this process—their violence reflects 'who they are' (barbarians, colonizers, terrorists), whereas our violence is shrouded in circumstance ("we had to kill them because …").

The neural infrastructure built up around maintaining ideological righteousness is immense. Dozens of distinct biases have been identified, named, and characterized.[8] And since these processes occur automatically, in regions of our brain that are generally inaccessible to conscious introspection, we are subject to their effects whether we like it or not. We are, as the great psychologist Lee Ross said, "naïve realists" who believe that we alone see the world objectively, whereas those who disagree with us are inherently irrational. This "bias blind spot" is again not owned by some groups and not others—it is part of a consequence of having a human brain that is designed to operate efficiently.

The Brain on Violence

Finally, a hallmark of terrorism, in my definition, is indiscriminate violence against members of the "out-group." Inter-group violence is by no means limited to terrorist groups: Established governments and nation states have been responsible for the deaths of hundreds of thousands of people over the past decade. How different is "their" violence from "ours?"

Our brains are shaped with the capability to care deeply, but also to kill. This deep ambivalence is potentially problematic. A society filled with people who are inherently very compassionate and very violent might prove unstable. Part of evolution's solution to this problem seems to have been to tether the processes that undergird pro-sociality (e.g., empathy) and the processes that enable violence (e.g., dehumanization) to in-group and out-group distinctions. In this way, people would be potentiated to love the in-group and hate the out-group; to fight and die on behalf of "us" and to be willing to kill "them." The psychological processes that drive deep altruism (for the in-group) and motivate extreme violence (toward the out-group) still live within us.

In the US, we fight others by proxy with our professional militaries, so we are rarely put in a situation that would involve directly harming out-group members. But the willingness to harm others can still be assessed among nonmilitary citizens of Western democracies. Since experimental evidence is scant from actual members of terror groups, I will provide the evidence for group-based violence from "us"—members of mostly majority groups that have the potential to act violently on our behalf. I will argue that these processes are similar to those acting in terrorists who actually pull the trigger.

So what drives someone to commit political violence (ideologically motivated violence), more generally? I find it useful to think of the psychology of political violence as a collection of impulses within us that tug us either toward or away from violence. If the various pulls toward violence are strong enough and the pulls away from it weak enough, a person engages in political violence; if not, they don't. Below is a brief outline of work I've done to illuminate two of the processes contributing to this psychological calculus: empathy and dehumanization.

The Empathy Factor

We are accustomed to thinking of empathy as an unambiguous force for social good. And for sound reasons—empathy is a "social glue" that arguably has been fundamental in enabling large groups of unrelated humans to band together in complex, cooperative societies. Although good experimental evidence shows that the amount of empathy one possesses (i.e., trait empathy) or expresses (i.e., state empathy) can drive altruism, there is also reason to believe that empathy may not be as unambiguously pro-social in intergroup contexts.[9,10] Specifically, whereas empathy for an out-group likely motivates pro-sociality toward its members, in-group empathy may have the opposite effect: If people feel the suffering of in-group members particularly acutely, this may motivate them to act against members of an out-group that they see as responsible.

In experimental research, I have tested the effects of in-group empathy and out-group empathy (and the difference between the two; i.e., "parochial empathy") in three contexts: Americans regarding Arabs, Greeks regarding Germans (during the Greek financial crisis), and Hungarians regarding Muslim refugees (during the refugee crisis). Predictably, in all of these settings, the more empathy participants reported feeling for the suffering of random out-group members, the greater their willingness to help and the less their willingness to harm needy members of that group (e.g., donations to civilian victims of drone strikes). However, empathy for in-group suffering predicted the opposite: less willingness to help the out-group and more willingness to harm. In fact, this is the conclusion drawn by a number of researchers who have interviewed attempted suicide bombers or families of people who had engaged in suicide bombings. Although some who commit political violence appear to be unhampered by empathy, the majority tend to be characterized by a strong communal focus that includes compassion and caring for others.[11]

Empathy therefore contributes two ropes to the internal tug-of-war: The greater the pull from in-group empathy to harm the out-group, and the weaker the pull from out-group empathy to prevent this, the stronger the overall motivation to engage in or condone intergroup aggression. It is therefore the difference in empathy, rather than the capacity for empathy, that best predicts intergroup violence.

The Dehumanization Factor

Historically, dehumanization has accompanied some of the darkest chapters in human history. During colonization, slavery, genocide, and war, depictions of the other side as uncivilized brutes or animals has been commonplace. We see this type of dehumanizing rhetoric from terror groups today—not only are we, the "infidels," referred to as "pigs" or "dogs," but we are viewed as undifferentiated and therefore collectively responsible. The rhetoric in Western democracies about disliked Muslim groups and terrorists is nearly identical: Iranians, Hamas, and ISIS have been depicted in the mainstream media as rats, beasts, snakes, or vermin in need of extermination.

In recent work, I have attempted to go beyond current psychological trends that use subtle measures of "everyday dehumanization" to capture overt expressions of dehumanization that were typical of colonial times and still seem present today. Toward this end, we developed a measure that captures blatant dehumanization using the popular "Ascent of Man" diagram, which depicts evolutionary "progress" with five images, from a quadrupedal early human ancestor through fully upright "modern man." By asking people to indicate where on the image certain groups fall, we have been able to assess levels of perceived "humanity" among a range of participant groups, towards a host of targets.[12]

To our academic delight (and personal dismay), we have found that people from every country we have assayed (the US, England, Denmark, the Netherlands, Spain, Greece, Hungary, Israel, Jordan, and the state of Palestine) rate at least one other group to be at least 15 points lower on the 100-point Ascent dehumanization scale than their own.

At the individual level, ratings of Ascent dehumanization are highly consequential. In Europe, for example, the degree to which people dehumanize Muslim refugees predicts their support for antirefugee policies and resistance to refugee settlement, even when accounting for conservatism and prejudice.[13] In the US, levels of Ascent dehumanization are associated with positions on a range of socially relevant issues, including willingness to sign petitions opposing the Iranian Nuclear Accord. In a recent study inspired by anti-Muslim rhetoric from the presidential campaign, we found that the dehumanization of Muslims was strongly associated with the willingness to punish all Muslims for individual acts of terrorism.[14]

What's more, the consequences of dehumanization go beyond how they motivate members of the dehumanizing group—they also affect the dehumanized. Specifically, we found that the more dehumanized Muslim Americans *feel*, the more likely they are to dehumanize non Muslim Americans, which leads to greater support for violent forms of collective action (i.e., "by any means necessary)," and less willingness to report suspicious activity in their communities to the FBI.[15] Making others feel dehumanized therefore puts us all at greater risk of that group allowing violence to happen, which could be interpreted by the dehumanizing group to justify (and compound) their dehumanization. Of course, the interaction between meta-dehumanization, dehumanization, and support for violence can easily ratchet up intergroup conflicts.

Thus, if dehumanization cuts a psychological thread that normally inhibits intergroup aggression, "meta-dehumanization" can provide the scissors. It is certainly not much of a stretch to imagine that terrorists think that they are dehumanized by Westerners. In fact, given the prevalence of dehumanizing depictions and language used to describe terrorists, it would be shocking if they did not.

Pondering the Future

While we like to think of "terrorists" as sociopaths and misfits distinct from "us" and united with each other by shared pathology and unfettered hatred, in fact their most salient characteristics—fervent attachment to a group ideology and a willingness to engage in indiscriminate violence—are likely driven by deep psychological processes shaped in the human mind through evolution. The great irony, then, may be that the best way to understand the mind of a terrorist is by examining our own.

Much of how we view terrorists is built upon a series of assumptions. Primary among these is that "they started it." Afghanistan and Iraq were a *response* to 9/11. But from terror groups' perspective, they are the aggrieved party and it is the other side that started it. An insightful observation recently came from a special forces lieutenant who related overhearing some of his men talking about the Taliban they were fighting in Afghanistan. Referring to the classic patriotic film *Red Dawn*, where a group of rural Americans fight off the invading Russian army, one of the American soldiers said to the other: "If this is *Red Dawn*, we're the Russians." If we accept that *they* are right and we attacked them first (even if only in their own minds), then how differently do we imagine we would behave in their situation?

The insightful comments from the soldiers quoted above notwithstanding, the reality is that we are using Stone Age psychology to solve 21st century conflicts. But there is hope. As much as evolution has baked into the human brain psychological processes that lead us stumbling into conflict, that same brain is endowed with an overriding organizational principle: flexibility. As powerful as these destructive unconscious forces may be, we are built to be able to gain conscious control of them. For example, our recent work has shown that if people are made aware of the hypocrisy of holding Muslims as a group responsible for terror attacks without considering white people similarly responsible for violence committed by white supremacists, their collective blame of Muslims dramatically decreases, which then ameliorates their endorsement of violence against Muslims.[16] Determining which of the unconscious biases that underlie our ideological certainty can be inoculated against is one step toward mastering the destructive tendencies in our own minds.

One of the great gifts that science has given to humanity over the past 2,000 years is humility. The Earth is not the center of the universe. Our DNA is not fundamentally different from that of other living things. And our brains do not differ markedly from those we fear or hated others. The great hope from the neuroscience revolution is that awareness of our own brain may actually allow us to transcend the unconscious processes that drive us to conflict.

12

Finding the Hurt in Pain

By Irene Tracey, Ph.D.

Irene Tracey, Ph.D., holds the Nuffield Chair of Anaesthetic Science and is head of the Nuffield Department of Clinical Neurosciences at the University of Oxford. Tracey did her undergraduate and graduate studies at the University of Oxford and then held a postdoctoral fellowship at Harvard Medical School. She helped to co-found and for ten years was director of the Oxford Centre for Functional Magnetic Resonance Imaging of the Brain at the university. She was an elected councillor to the International Association for the Study of Pain and chair of their Scientific Program Committee, is a Council member of the Medical Research Council, and is on the Brain Prize selection committee. In 2008 she was awarded the Patrick Wall Medal from the Royal College of Anaesthetists and in 2009 was made a Fellow of the Royal College of Anaesthetists. In 2015 she was elected a Fellow of the Academy of Medical Sciences and in 2017 was awarded the Feldberg Prize. Tracey is married to Professor Myles Allen, a climate physicist, and they have three children.

Pain is unique to every person, and difficult to quantify and treat. Whether it is delivered as a jolt or a persistent, dull ache, pain is guaranteed to affect one's quality of life. Our author examines how brain imaging is opening our eyes to the richness and complexity of the pain experience, giving us extraordinary insight into the neurochemistry, network activity, wiring, and structures relevant to producing and modulating painful experiences in all their various guises.

A WISE MAN, PURPORTED TO BE OSCAR WILDE, once said: "I don't mind pain, so long as it doesn't hurt." Packed into that flippant comment may be insight beyond his intent.

The ability to experience pain is shared across species. Acute pain is the body's alarm and warning system and, as such, a good thing. It is key to survival, and is evolutionarily old. All living things have the ability to detect factors in the environment that might "hurt" and cause injury, harm, and, ultimately, perhaps death. These factors are lumped into thermal, chemical, and mechanical categories. Without pain you simply don't survive. We know this, sadly, because a rare genetic condition, congenital insensitivity to pain, produces a phenotype in people such that they don't get the "good" warning pain after damaging themselves; historically, they didn't survive to adulthood due to the consequences of unfelt injury.

Why does congenital insensitivity to pain cause death and why does experiencing pain aid survival? One reason is that pain motivates decisions to act. Think about an everyday painful experience: One picks up something hotter than expected. One's options are drop it and make a mess or grin and bear it until a solution is found. In an instant one has detected it's hot (thermal), it's on the hand (location), it's painful (intensity), one doesn't like it (unpleasant), attention is now fully directed toward it (cognition), and one is unhappy about it (emotional). But what is one going to do?

Based upon learned responses, past experiences, and competing other interests (being told off for dropping it), we make a decision and act. Recruiting extraordinary brain-based networks, we're able to block the pain and get the hot item safely to a place where we can set it down. Then we act again to nurse our injury—perhaps running our hand under cold water. For the person who cannot feel pain, there is no warning sign and the injury, perhaps considerably more than the one described, might produce an undetected infection and subsequently death. So, pain is essential for survival.

The nature of pain, however, is hard to understand, and pin down, and is, in a word, complex. Pain, by definition, is a subjective and private multidimensional experience. Further, it's highly malleable, depending upon the context, and the cognitive and/or emotional state in which we experience the injury. In short, subjective pain is not linearly or simply related to the tissue-damaging signal input. The classic soldier on the battlefield or sportsperson not experiencing pain during an injurious tackle on a rugby pitch are the exemplars illustrating how evolution has fine-tuned our nervous system to be capable of changing the pain experience we have. Why? Because this means we'll make the right decision about how to act in response to that pain. For the interested reader, there are several excellent books that discuss pain and its importance to life and influence on history and society that I recommend.[1-5]

Understanding Signals

Confronted by a lion, we can make the decision to scream in agony after he or she has bitten our arm off or block the pain from this injury so that we are better able to run away. Similarly, if we can modulate the incoming signals from the damaged tissue (nociception) along its journey to the brain, from which the conscious experience of pain emerges, then perhaps we can generate a pain-like experience without a nociceptive input using brain-based networks when perceiving a threat of pain. Historically, if people reported pain in the absence of identifiable tissue injury, it was called "psychogenic" pain. Back in the day, this was a pejorative term because there was little understanding of the mechanisms that underpin nociceptive processing and pain generation by the peripheral and central nervous system. The same was true for patients having undergone a placebo test to "catch them out.'"

Now we know better and such terms are respected. Indeed, the very definition of pain by the International Association for the Study of Pain (IASP) allows for pain occurring with and without tissue damage: *"An unpleasant sensory and emotional experience associated with actual or potential tissue damage, or described in terms of such damage."* The IASP goes on to say that an inability to communicate verbally does not negate the possibility that an individual is experiencing pain and is in need of appropriate pain-relieving treatment. Individuals learn the application of the word "pain" through experiences related to injury in early life. As such, we must always trust what they say about *their* pain, irrespective of what looks like a similar injury because this will be felt and experienced differently for these and other reasons, such as genetic and epigenetic influences. Pain is not a unitary thing and no two pains are the same, even in the same individual.

Despite this complexity, can we do better in understanding why someone's pain is the way it is? What about the demented elderly, comatose individuals, anaesthetized patients, or nonverbal infants who don't have speaking skills or options to describe their pain? Brain imaging has been one such tool that has opened our eyes to the richness and complexity of the pain experience, giving us extraordinary insight into the neurochemistry, network activity, wiring, and structures relevant to producing and modulating painful experiences in all their various guises. Techniques such as functional, diffusion, and structural magnetic resonance imaging; positron emission tomography; and electro- and magneto-encephalography are now widely used to understand acute and chronic pain.[6-8]

The High Costs of Chronic Pain

Chronic pain is defined as pain that persists beyond normal tissue healing time. It is estimated that one in four adults has a persistent pain state that, on average, lasts approximately seven years (20 percent for more than 20 years). As such, it brings considerable suffering to patients and their families, alongside significant costs to society (estimated at 200 billion Euros annually in Europe and $630 billion annually in the US). Co-morbid problems like depression, anxiety, and sleeplessness are inherent in chronic pain.[9]

Sadly, current treatment options don't provide adequate relief to the majority of patients. As such, it's one of the largest medical health problems worldwide. But scientific research at a preclinical and clinical level using an array of techniques—from molecular and cellular biology to advanced neuroimaging—is giving us unprecedented insights into chronic-pain states. A paradigm shift in our thinking has occurred, one that is slowly unraveling throughout the medical and scientific communities.

We've stopped thinking about chronic pain as a continuation of what caused the initial, perhaps acute, pain. Chronic pain is a whole new state, with its own underpinning mechanisms that can be shared across many different types of chronic pain despite completely different initiating causes (e.g., symptoms are similar whether it is painful diabetic neuropathy and neuropathic pain states caused by traumatic nerve injury or having chemotherapy—and this means that the underpinning mechanisms must be similar, too)—almost considering chronic pain now as a disease in its own right. This new way of thinking has given us insight into a new biology with new mechanisms to target, making the future very bright for chronic-pain sufferers.[10]

Getting to the Roots of Pain

Now, back to the basic neuroscience question of where the "hurt" is of pain. In short, we still don't know, but we're closer. What we have learned after 20 years of neuroimaging is that the brain is key to experiencing pain—not impressive when you consider that Hippocrates suggested this was the case thousands of years before we were certain that the brain was the organ for perception and sensation.

We know that the brain responds to "painful" or nociceptive events in a host of brain regions spanning sensory, discriminatory, affective, emotional, cognitive, brainstem modulatory, motor, and decision-making circuits in a flexibly accessible manner. Not all regions within this expansive network activate every time, even to the same nociceptive input or injury, and certainly not to the same extent. This ability to activate a varying set of brain regions in a highly flexible manner provides for the endless possibilities of varying painful experiences people need if they are to have "the pain that is appropriate for the situation they are in," such that they mount the right decision and action as to what to do about it.[11]

Imaging reveals that many of these brain regions are not pain-specific but nonetheless relevant for providing that rich multidimensional experience that is pain (e.g., threat, fear, attention networks). My group and others have used novel methods and paradigms to help dissect neuroanatomically this complex network in order to disambiguate which brain regions subserve the different features of a painful experience.[12]

Using psychological paradigms, pharmacological agents, novel imaging methods, and various injury models of peripheral and central "sensitisation" or amplification, we've been able to relate neurophysiologic measures from advanced brain imaging to perceptual or nonperceptual changes in pain experiences induced by these methods. Noninvasive identification of where functional and structural plasticity, sensitization, and other amplification or attenuation processes occur along the pain neuraxis for an individual, and relating these neural mechanisms to specific pain experiences (measures of pain relief, persistence of pain states, degree of injury, and the subject's underlying genetics), has neuroscientific and potential diagnostic relevance. As such, advanced neuroimaging methods can powerfully aid in explaining a person's multidimensional pain experience, analgesia, and even what makes them vulnerable to developing chronic pain.[13]

Let me illustrate with examples. People report that when they feel sad, their pain is worse. Does sadness influence the physiological processing to change the experience? Understanding this has relevance to our understanding and treatment of depression in pain as a significant factor, as well as changing attitudes/biases/myths toward pain.

In a test performed by my laboratory to explore how mood and pain interact, we played healthy students Prokofiev's *Russia under the Mongolian Yoke* at half-speed as they read negative statements ("I have no friends," "My life is a failure," and so forth) and were given painful stimuli. Perhaps unsurprisingly, compared to the control mood condition (listening to Dvorák and reading neutral statements), the subjects rated the same stimulus in the two mood conditions as more painful when they were sad. Subtracting brain activity produced to the stimulus during the neutral mood from that produced during the sad-mood condition confirmed that more activity in various brain regions (sensory and affective) occurred and accounted for the heightened pain reported during sad music. In short, normal emotion regulatory circuitry was disrupted in the sad condition, and this influenced how pain was processed within the brain to increase its activity levels in various regions (e.g., amygdala, insulae, inferior frontal gyrus, anterior cingulate).[14]

Therefore, sadness produces a physiological amplification—akin to a volume button on a hi-fi—by brain-based mechanisms. Related experiments simulating going to the dentist and being anxious, terrified, and threatened have also been done by my group and others. Again, the story is clear. Manipulating a healthy subject's emotional state negatively (i.e., make them anxious or threatened) changes how they perceive the very same painful stimulus toward being more painful. The 'anxiety volume' button, if you like, appears to be centered on the hippocampus/entorhinal complex with interactions to the anterior insula and mid anterior cingulate.[15,16]

Understanding Clinical Pain

These basic findings have proved useful to interpret and understand clinical pain. Knowing that such regions are involved in a patient's painful experience highlights the relevance of treating these co-morbid factors just as seriously as the inciting nociceptive input. The medical model likes to see tissue damage in order to believe a person's pain. But now we know that factors such as sadness, anxiety, and threat can amplify nociceptive inputs via such neural processing circuits and make the pain worse. This helps us to understand why there is often a mismatch between what is observed in terms of injury and what the patient or person reports is his or her pain.

Also to be considered is that sometimes pain isn't so severe as it appears. Remember that circus trick involving lying on a bed of nails or walking across hot coals? People can be distracted from their pain by, say, listening to music or watching a gripping film. Experiments have shown that when you are distracted from pain, an evolutionarily old system centered in the brainstem and driven by subcortical and frontal cortical regions, and which is shared across species and unique to the pain system, is recruited to drive descending inhibition and block nociceptive inputs arriving into the spinal cord from the injured body part.

Consequently, there is less input to the brain, so less pain. This system, called the Descending Pain Modulatory System, has unfortunately an opposite and facilitatory action that makes pain worse. Current work suggests that an imbalance between this inhibitory and facilitatory action is important in chronic pain. This is why soldiers on the battlefield don't feel pain when their attention is diverted.[16-18]

Naturally, the question then asked is whether placebo analgesia uses the same brainstem mechanism or is something different. Placebos have a checkered history born from ignorance. The word placebo describes the Latin chants sung at funerals by hired mourners, and because of this history its use invokes feelings of fakery, deception, and lies. To get a placebo effect requires conditioning or having someone learn to expect a certain outcome from a particular ritual that might be a treatment in a medical setting. It was once assumed that if the person had a placebo effect then he or she was faking it or lying, and placebo tests were used for such detection purposes. But as far back as Hippocrates' and Galen's time, it was known that physicians caring for patients could produce remarkable healing without any pharmacological aids. We now have proof that these effects are actual physiological changes in the body that can be recruited and activated by the mere expectation of an outcome.[19]

The Role Placebos Play

Neuroimaging helped prove the mechanism behind placebo analgesia. It has shown that, for the most part, it is simply expectation driving the descending pain (inhibitory) modulatory pathway from the brainstem to the spinal cord, such that as less nociception (the nervous system's response to certain harmful or potentially harmful stimuli) arrives to the brain, less pain results. Of course, nocebo effects can be produced where pain is made worse, and imaging has also proved useful in identifying its neural basis.[20,21]

Why is this important? Well, to know that the brain has a powerful system for modulating the physical experience that a patient has—in this instance his or her pain condition for better or worse—again helps us understand the nature of a patient's pain. A patient may even set and manage his or her own pain expectations based on what he or she has Googled. Knowing and managing what his or her expectations may be can contribute to actual treatment outcome.

We simulated this precise scenario in an experiment where healthy subjects were given a powerful intravenous opioid (painkiller) during a brain-imaging study throughout which we gave them painful stimuli. We manipulated their expectation of having this analgesic drug by simply not telling them when we started the infusion (hidden injection), then telling them we were starting it (driving positive expectation—even though they were already now on the drug), and pretending we stopped it when we didn't (driving negative expectation or nocebo). The results were striking. There was a small analgesic effect to the painful stimuli when the drug was given by hidden injection; we doubled that analgesic effect by driving positive expectation (even though nothing changed in terms of the drug dose being delivered), and then overrode and killed all the analgesic effect of the opioid when we pretended we stopped the infusion (which we didn't—so subjects were still on the drug but thought they were not), returning their pain ratings to preinfusion levels of the drug.[22]

The imaging revealed how these effects were produced and confirmed that the descending pain modulatory system was recruited during positive expectation and the network by which anxiety makes pain worse (the hippocampal neural "amplifier" of pain discussed above) was recruited during the nocebo part when they thought the infusion had been stopped and their pain ratings returned to preinfusion levels! This experiment speaks to the value of physicians and patients having time to discuss their condition and treatment options, so that expectations can be reliably managed and not unduly influence treatment outcomes.

Hedonic Flipping—Making Pain Pleasant

Is it possible to not just block pain, but actually make pain pleasant? English philosopher Jeremy Bentham once suggested that we seek pleasure and avoid pain. But that's a relative judgment. Sometimes pain might be the lesser of another evil, in which case its subjective value can be pleasurable or thought of in such terms. There are everyday examples where reappraising pain as something pleasurable or rewarding is done—and not by the sado-masochist. After a vigorous run, for example, runners may not perceive their tired body or aching muscles as unpleasant pain. Rather, they may perceive what they feel as pleasant pain, indicating health, exercise, and achieving their goals.

Imaging by our group has explored the neural basis of "making pain pleasant" and how that might be harnessed in a clinical setting. Changing the hedonic value of pain is achieved by the reward system and the descending modulatory pain system working in concert to reduce the intensity of the painful inputs and make their value appear rewarding.[23,24]

The Pros and Cons of Brain Reading

Finally, a new area of imaging is emerging that has the potential to impact other spheres of society beyond the laboratory and clinic: brain reading. For pain, such an approach might be useful in situations where pain cannot be communicated through speech or behavior (i.e., an infant, a demented elderly person, or a comatose or anaesthetized patient). The approach is to use neuroimaging alongside algorithms that are trained to classify patterns of brain activity, or "brain reading," based upon what we know the person thought or felt during that pattern generation. Then, classifier algorithms predict what experience the person had by putting his or her brain data during that experience into a classification computer program or multivariate pattern analysis.

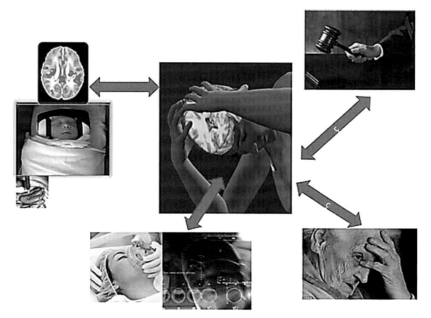

Various situations where a neuroimaging-based 'brain reading' approach might be useful.

The accuracy has proved reliable and the approach popular with companies selling the technology to insurance companies, courts of law, and situations where there is a desire to have presumed independent and objective readouts. But the practice is controversial. Complex interpretative issues need to be resolved before brain reading is used as a guide in determining an individual's pain.[25,26]

What the next decade brings about in perceptions of pain will hopefully generate benefits to patients and society at large, as well as enrich our neuroscientific knowledge regarding this most complex of subjective experiences.

BOOK REVIEWS

13

Review: John Seamon's Memory & Movies: What Films Can Teach Us about Memory

By Alan A. Stone, M.D.

Alan A. Stone, M.D., is the Touroff-Glueck Professor of Law and Psychiatry in the Faculty of Law and the Faculty of Medicine, Harvard University. He has been a Guggenheim Fellow, a fellow at the Center for Advanced Study in Behavioral Sciences, and the Tanner Lecturer at Stanford University. At Harvard, he has been a fellow of the Mind Brain and Behavior Interfaculty. Stone has also served as president of the American Psychiatric Association. Stone is a graduate of Harvard University (1950) and Yale Medical School (1955), is the author of several books, and reviews films for the *Boston Review*.

From trauma to amnesia to senior moments, memory has been a major plot line in films since the 1942 classic Random Harvest. *John Seamon, an author and professor of psychology whose research includes how a camera aids memory and the impact of storytelling on memory, has shifted his lens to focus on how memory has been portrayed in one of the world's most beloved art forms.*

IN LUCID PROSE, John Seamon has explained almost everything we know about the several systems that constitute memory and has enlivened the account with illustrative examples from 40 well-chosen films. It is a polished performance, aimed at a general readership.

The modern marriage of memory science and movies was launched by Christopher Nolan's film *Memento* (2000), an account of a man with severe anterograde amnesia. The film is a complex, neo-noir puzzle told backward and forward simultaneously. It is about murder, memory, grief, revenge, and identity, and conveys the disorientation that Leonard, the disabled protagonist, is experiencing. Rejected as too complicated by the major film distributors, *Memento* found its own audience in art house cinemas around the world, earning cult status and commercial success. Few films have provoked so much discussion and dissection by critics, film buffs, and neuroscientists, who tried to solve the puzzle of what actually happened.

Seamon, a psychology professor at Wesleyan University, was inspired by all this to try an experiment. He asked students in his memory class to watch *Memento* and then write an essay describing what the filmmaker got right and wrong. The experiment was a great success; his students "responded enthusiastically and said they saw the film in a new light."

Something should be said about this "new light." Based on 20 years of experience using films in law school classes to explore the nexus of psychology and morality, I offer a possible explanation: Film is the medium of young adults and they feel confident about their judgments and opinions. Furthermore, by inviting his students to critique the film, Seamon transformed what is ordinarily a passive mind-set—entertain me—into an active mind-set of intellectual engagement. Like all good teachers, he empowered his students. Out of this experiment came a successful course in memory and the movies, and this highly readable book.

Seamon's book is subtitled *What Films Can Teach Us about Memory*. This seems a dubious claim. Films by themselves have taught us very little about memory. In fact, as Seamon emphasizes, filmmakers for most of the 20th century were misrepresenting amnesia as a plot device. Typically, a bump on the head would lead to retrograde amnesia characterized by loss of identity. This would be cured by another fortuitous bump. As Hollywood came under Freud's influence, the physical bump on the head was replaced by psychological trauma.

Even the *Memento* craze did not deter Hollywood from relying on the traditional plot device of retrograde amnesia with loss of personal identity. In 2002 came *The Bourne Identity,* about an amnestic Central Intelligence Agency assassin; it had three sequels, and together the *Bourne* movies grossed over a billion dollars worldwide. Seamon, in his concluding chapter, asks: "Why is amnesia so often misrepresented in film?" Perhaps because fiction is more profitable. And although they misrepresent how the mind-brain remembers, these films touch on the mystery of the self that still challenges science and philosophy.

Seamon begins his book with a discussion of the romantic comedy *50 First Dates*, starring Drew Barrymore as Lucy. Lucy's memory is impaired after a car crash. She can remember her past and each day as it unfolds, but cannot retain the new memory—hence the 50 first dates. Of course this kind of amnesia, a great plot device for a comedy, conforms to nothing we know about how memory actually works. Seamon then recounts the case history of a woman, a superfan of Drew Barrymore, who a year after *50 First Dates* had an automobile accident and supposedly developed the same kind of amnesia on a functional/psychological basis. What should we make of this apparent amnesia by identification?

Seamon provides no answer about this particular case, and says little throughout the book about functional impairments of memory. His equivocal "take home message" from this life-imitates-art example "is not that movies are an inherently misleading way to learn about memory. ... [we must]... let our viewing be guided by scientific knowledge." In other words, without Seamon's guidance, films would teach us precious little. What he proceeds to do in the book is to use films as specimens that he summarizes and dissects with scientific expertise, demonstrating what they get right and what they get wrong.

Along the way, in concise chapters (each begins with "setting the scene" and ends with a "fadeout"), he illustrates the science of memory: working memory, lasting memory, recognizing faces, autobiographical memories, persistent traumatic memories, the reality of amnesia, senior moments and Alzheimer's disease. Considering that this is a popular book, Seamon has been scrupulous in citing the literature for every statement he makes.

Particularly impressive is the way he navigates the treacherous waters in the chapter "When Troubling Memories Persist." Clint Eastwood's *Mystic River* and Eugene Jarecki's documentary *Capturing the Friedmans* are featured in his discussion of childhood sexual abuse and the phenomenon of recovered memories. Clinicians who are still fighting the "memory wars" and the reality of "Freudian repression" will find Seamon both objective and informative.

Not that he gets everything right in this book. In his otherwise excellent chapter on the "reality of amnesia," he briefly comments on electro-convulsive therapy (ECT), once a controversial treatment for depression. ECT is now widely accepted by clinicians as highly efficacious when antidepressant medications fail. Seamon misleadingly presents it as a stopgap measure that causes patients "to forget troublesome issues."

One of the problems in a book about film is that readers who are moviegoers can be fanatically indignant about spoilers. Seamon seems blissfully unaware of the issue as he neatly summarizes the plots of his films. Cinephiles might also find fault with summaries that, by emphasizing the theme of memory, seem to miss the real meaning of the film. For example, in Seamon's summary of the award-winning *Life of Pi*, he briefly describes the two interpretations of what happened on the life raft after the shipwreck, omitting any mention of the film's deeper meaning as an exploration of faith.

But these are quibbles. Seamon has produced, directed, written, and narrated a book that will educate as it entertains. It might even empower its readers.

Seamon also plans a free online course on the book.

14

Review:
David Casarett's *Stoned:*
A Doctor's Case for
Medical Marijuana

By Bradley E. Alger, Ph.D.

Bradley E. Alger, Ph.D., is a professor emeritus in the Department of Physiology at the University of Maryland School of Medicine. He received his Ph.D. in experimental psychology from Harvard University in 1977, and taught and did research at Maryland from 1981 to 2013. In the early 1990s, Alger and Thomas Pitler characterized the first signaling process ultimately found to be mediated by endocannabinoids in the brain. Alger has authored over 120 research papers and reviews, focusing in the past two decades on the regulation of synaptic inhibition and endocannabinoids.

With legal cannabis sales at $5.4 billion in 2015 and expected to rise by another billion in the United States in 2016, legalization of marijuana and its impact on health is a hot topic of national debate. Casarett, a physician at the University of Pennsylvania, immerses himself in the culture, science, and smoke of medical marijuana in order to sort out the truth behind the buzz. Our reviewer, who has authored more than 120 research papers and reviews on the regulation of synaptic inhibition and endocannabinoids, tells us what the author got right, but also overlooked on his journey to learn more about a complex and controversial subject.

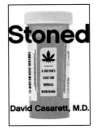

DAVID CASARETT WAS A palliative care doctor with an Archie-Bunkeresque level of skepticism regarding medical marijuana. He doubted that marijuana was a medicine, or indeed that it was good for anything, but finally had to admit that he didn't know enough to advise patients who asked about it. Does marijuana "work"? Is it safe? Effective?

This book chronicles Casarett's foray into the world of medical marijuana. It is an engaging, lively, thought-provoking tour seen from the street, not the laboratory; the walk-in clinic, not the ivory tower. The doctor wants to know not only the subject but also how to explain it to his patients (and readers) in terms that they will understand; how to give them a voice in their own care and be informed medical consumers. In trying to accomplish this, he covers a lot of ground.

Casarett discusses a range of maladies for which marijuana is said to be beneficial—including insomnia, nausea, cachexia, pain, and cancer—in vignettes that begin with an arresting anecdote or personal story of a patient (including himself in one case). He establishes a largely jargon-free scientific/medical context for understanding how marijuana might act in a given case, and sums up his impressions of the evidence. This is advice such as you'd get from a neighbor (who happens to be a doctor) over a beer after a game of golf: many "possibles" and "maybes," a few numbers, but no charts and graphs, and only a couple of firm answers.

The uncertainty and caveats are unsurprising because many of the experimental studies available are small and not well controlled. Marijuana "seems to be" effective in treating neuropathic pain, "definitely" works for nausea, "probably" improves appetite, and reduces insomnia; it "might be" helpful for anxiety and post traumatic stress disorder; "maybe, someday" we'll know whether it does anything for cancer, but now, nothing. The reader, who may be frustrated by the indefiniteness of his verdicts, is reminded that the scarcity of hard data results from the benighted federal drug policy that still classifies marijuana as a Schedule I drug (dangerous and of no medical value), significantly worse than morphine, cocaine, or amphetamines, which are on the less restrictive Schedule II.

Usually Casarett gives us enough scientific background to clarify his opinions without overdoing it. His accounts of why marijuana affects different people differently, and how the storage of THC (the psychoactive chemical in cannabis) in body fat can modulate its effects, are two good examples among many. But the book is as much sociology as medicine. Casarett often goes undercover to capture the experience of the individual patient peering in at the medical marijuana subculture. At one point he gets a tutorial in the psychoactive subtleties of marijuana varieties that is as nuanced as the wine recommendations of a sommelier at a tony New York restaurant (Casarett takes it all in, but is a noncustomer).

He reviews the panoply of forms and delivery methods of cannabis products—besides the standard joints, there are pills, vaporizers ("vape pens"), oils, resins, oral sprays, potables (cannabis-infused beer and wine), and edibles from gummy bears to brownies—and weighs their pros and cons. He recounts his own attempt to treat chronic back pain by smoking a joint on his back porch: It is neither transformative nor a complete nightmare, although one doubts that he'll go there again. He does answer a commonly asked question: Why smoke if you can get cannabinoids in FDA-approved pills, or edibles? In a nutshell: control. Because of the rapid transit time for THC to go from the lungs to the brain (tens of seconds), an experienced user can titrate his intake to produce just the desired level of symptomatic relief. Taken by mouth, THC has to pass through the GI tract (tens of minutes, with times dependent on what food was eaten and when, etc.) and undergo variable absorption into the bloodstream; no wonder the effects of ingested marijuana are less predictable. Couple this lack of control with the disinclination of severely nauseated patients to swallow anything, and one appreciates the appeal of smoking.

Despite the book's subtitle (*A Doctor's Case for Medical Marijuana*) this is not a tale of advocacy; the author shuttles evenly between doubt and sympathy. A hilarious visit to a sketchy marijuana clinic/dispensary that will confirm the worst suspicions of die-hard opponents who see the entire medical marijuana movement as a scam, is counterbalanced by moving stories of people who, having tried conventional medications (including morphine) without success, depend on the comfort that they get from marijuana to live a normal life.

Casarett's authorial instinct for the captivating image occasionally leads him astray: He repeats a story of some cannabis-dependent soldiers in the 1940s and their lurid and sometimes violent behavior when compelled to go cold-turkey during assignment to a cannabis-free environment. This anecdote, seemingly right out of the *Reefer Madness* handbook, is used to dramatize the withdrawal symptoms that might accompany cessation of marijuana use, although Casarett acknowledges that this case is atypical (and hardly a controlled study). He is alarmed that 9 percent of marijuana users meet the clinical definition of addiction (as compared with 12 percent of alcohol users and 15 percent of heroin users), and takes it as a given that any addiction is bad.

The discussion would have benefited from a more critical analysis. For instance, given the numbers, shouldn't we promote marijuana use as a way of reducing the overall heroin addiction rate? Or consider what Casarett doesn't stress: that overdoses of opiates or alcohol are often fatal. In 2014 opiates caused 25,000 deaths (DrugAbuse.gov), and alcohol-poisoning causes 2,200 deaths each year (CDC website), whereas, as Casarett notes, deaths from marijuana overdose are essentially unknown (DrugWarFact). Finally, alcohol consumption was implicated in 10,076 deaths from car crashes in 2013 (CDC website). Despite the presence of millions of recreational users in the US, there is no evidence that marijuana causes anything like that level of carnage. Nobody is recommending marijuana use as a public health safety measure—you shouldn't operate cars or heavy machinery when stoned—but these are some of the societal complexities that the book skirts.

Given his cautious conclusion that marijuana can be beneficial in some instances, it may come as a surprise that Casarett is not bullish on marijuana's future as a medicine (he considers it an "herbal remedy"), arguing that major pharmaceutical companies are working overtime to find drugs that will be better at treating the disorders that medical marijuana treats, and will not have marijuana's side effects. He cites the case of glaucoma, for which marijuana used to be recommended, but which is now controlled effectively by conventional medications. On the other hand, the discovery of the opioid receptor many years ago prompted confident predictions that opiate drugs would soon be available that would selectively relieve pain without causing euphoric or addictive sideeffects. The current epidemic of prescription opiate-drug addiction (and rebound heroin use) in the US is enough to give one pause. Will Big Pharma have better luck in replacing marijuana?

Casarett's engrossing narrative stance, basically as a physician playing the role of educated layman, perhaps leads him to overemphasize the interactions of the chemicals in marijuana; e.g., THC, CBD (a nonpsychoactive extract) with the major cannabinoid receptors, CB1 and CB2, for understanding marijuana's actions. Different drug-receptor interactions do contribute to marijuana's assortment of behavioral effects, but this narrow focus fosters the misperception that "the future of marijuana research" is in the hands of chemists who are tweaking the THC molecule and producing variants ("synthetic cannabinoids") that also activate the CB1/CB2 receptors. In fact, these variants will potentially interact with a large number of other molecular targets. As a case in point, anandamide, the classic natural CB1 activator ("endocannabinoid") in the body ("the THC inside all of us"), activates a noncannabinoid receptor, the TRP receptor, more efficiently than it does CB1! We will need to know much more about the molecular targets of synthetic cannabinoids before assigning them a leading role in medical marijuana-type therapies.

More significantly, Casarett skips over the myriad issues associated with the highly variable distribution of CB1 receptors across brain regions and functional classes of brain cells. Admittedly, this is a complicated subject, yet understanding it and figuring out how to target the cannabinoids correctly to carefully defined subregions will, I believe, ultimately be more relevant for developing marijuana-based therapies than refining drug-receptor match-ups. Finally, Casarett barely scratches the surface of the exploding field of the endocannabinoid system, exploitation of which will surely be a major direction for the future of medical marijuana. Why worry about exogenous cannabinoids if we can harness the ones we already have on board?

By and large, however, such lapses do not detract from my enthusiasm for the book. It accomplishes what it sets out to do, giving patients and caregivers a balanced, insightful view of medical marijuana in an entertaining, straight-talking way. I found it an enjoyable read and highly recommend it.

15

Review:
Santiago Ramón y Cajal's *Advice for a Young Investigator*

By *Michael L. Anderson, Ph.D.*

Michael L. Anderson, Ph.D., is associate professor of psychology at Franklin and Marshall College. His latest book, *After Phrenology: Neural Reuse and the Interactive Brain* (MIT Press), outlines a novel framework for understanding the evolution and functional organization of the brain. Anderson earned a B.S. in premedical studies from the University of Notre Dame, and a Ph.D. in philosophy from Yale University, and did his postdoctoral training in computer science at the University of Maryland, College Park. He was a 2012-13 Fellow at the Center for Advanced Study in the Behavioral Sciences at Stanford University.

Santiago Ramón y Cajal, a mythic figure in science and recognized as the father of modern anatomy and neurobiology, was largely responsible for the modern conception of the brain. The first to publish on the nervous system, he sought to educate the novice scientist about how he thought science should be done. We asked an accomplished young investigator to take a fresh look at this recently rediscovered classic, first published in 1897.

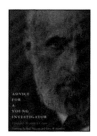

SANTIAGO RAMÓN Y CAJAL'S *Advice for a Young Investigator* might more accurately be titled *Advice for a Young Provincial Investigator*. Cajal, a renowned neuroanatomist who remains Spain's only Nobel Laureate in the sciences, wrote this and essentially all his works in turn of the century Madrid, and he acutely felt his distance from the capitals of European science. Paradoxically, however, this may well make the work more relevant to the young investigator in the globalized, interconnected world of today, for many obstacles to success are timeless struggles of the soul, and most of us are provincial in one sense or another. So, if you have just been elected a Junior Fellow of the Harvard Society of Fellows or have been anointed as brilliant, congratulations! There is little of use to you here. But if, like the vast majority, you hail from a rural university, a small liberal arts college, an underfunded hospital, or, yes, a struggling nation, then *Advice*, if not required reading, will at the very least repay the two afternoons it will take to absorb.

What is the advice of *Advice*? I distill seven themes below, but I'll also note that one of his important instructions is to learn from, but never trust, the work of commentators. I endorse that caveat; what will resonate in this rich volume will necessarily differ for each of its readers.

1. The only thing in your power is your preparation.

The role of chance in science is as undeniable as it is irreducible. You can do nothing to ensure success; there are no logical rules of discovery. But you can prepare yourself, and you must, so that when the unexpected finding, or odd phenomenon, or technological breakthrough occurs, it can be you who is best equipped to grasp its significance. Preparing means mastering bodies of knowledge and techniques known to be relevant, but also—and here I borrow not from *Advice* but from Cajal's remarkable biography—recognizing that sometimes knowledge and skills acquired from the necessity of circumstance will prove critical. So it was for Cajal, who, initially indifferent to medicine and needing a trade, was apprenticed to a barber. This granted such dexterity with a razor that when he finally came to the study of anatomy, his histological preparations were unusually skillful, enabling him able to see neural structures with exceptional clarity.

2. Be suspicious of "brilliance."

This theme is really a corollary of the first, but it leads to a different moral. Brilliance is mastery of skills and knowledge known to be important. It of course often leads to opportunity and success. But undue focus on this quality can engender worship of yesterday's abilities and insights, and this can hold individuals—indeed, entire fields—back. Never let perceived lack of ability limit you—*especially* when it's a self-perception. "Lack of ability" may simply be a set of skills waiting for its moment. Science needs a variety of minds. Besides, as Cajal writes, "work substitutes for talent, or better … it creates talent." The scientist's single most important virtue is perseverance.

3. Be appropriately respectful of authority, but no more.

No theory, no method, and no experimental paradigm is perfect. Do not defend or dismiss the errors of your teachers; use them to identify new problems to solve.

4. Balance concentration and relaxation.

Finding the time and space for extended and total concentration on a problem is vital. On these occasions shut out everything inessential: email, Facebook, politics. But when progress stalls after sustained and serious concentration, take a vacation. Plant a garden! The intellect thrives on a balanced combination of work and refreshment.

5. Don't worry about what it's good for.

All scientific findings are useful, eventually. The call to specify in advance the use of an investigation represents a lack of faith in the scientific enterprise, and of necessity limits creativity. Cajal lists a dozen phenomena that, had we waited to know what they might be good for before investigating further, would never have been good for anything at all, and we would now lack such tools as batteries, photography, and X-rays. Good science makes for good application, not the other way around. Are you listening, National Science Foundation?

6. Favor independence over resources.

When access to money, tools, or space would come at the cost of your autonomy, choose autonomy. Successful science can be pursued on any scale, but freedom of thought is nonnegotiable. This may seem naïve in the age of Big Science, but it reflects my own experience. My most important work was begun as an unfunded side-project and came to fruition in the lightly-subsidized lab of a small liberal arts college. My investigations were tailored to what undergraduate assistants could manage and, when I couldn't afford to generate my own data, I borrowed the data of others. You may object that such advice is hardly the basis for a universal principle: Those data took money to produce, just not my own. True enough. But this is advice to the young investigator. As the young mature, some will be successful enough to direct the large, well-funded projects that will produce the data, and miss the problems that will allow the next young investigators to move decisively beyond them.

7. Embrace scientific panglossianism.

This is the very best time to be a scientist. So was last century. And so will be the next. For there are no small problems in science, only phenomena so incompletely understood as to seem so. In science, notes Cajal (quoting Geoffroy Saint-Hilaire), "the infinite is always before us."

There is much more to this book than this sampling captures, including advice to the teacher, and even for choosing a life partner. That Cajal would think to include this latter is endearing, and, although his discussion is choked with objectionable stereotypes, the attempt reflects the positive and affirming thought that science is part of life, and should ideally be made seamless with it.

One of the strengths of *Advice* is its refusal to accept obstacles as excuses. This pull-yourself-up-by-your-bootstraps attitude will not please everyone—this is decidedly not a work of activism. (No bootstraps? Guess you'll need to make some first. No leather? How about twine?) But it is what gives the book its utility to the young investigator needing to make his or her way in their particular time, place, and situation. And the attitude is inspiring, in its way. Cajal was acutely aware not just of Spain's physical, but also its *cultural* distance from the great scientific centers of the world, and his call for Spain to acknowledge and repay its accruing "debt to civilization" has a Kennedy-esque ring. His book is intended in part, I think, as a reminder that those in a position to contribute to the advance of scientific knowledge, in whatever way and to whatever degree their situations and abilities allow, have a sacred obligation to do so.

And Cajal does mean sacred. It may be that the spiritual attitude that suffuses the book embodies the most important advice in *Advice*. Science is not just a collection of techniques or strategies for producing knowledge, it is an ethical stance of commitment to the truth. The good scientist knows that honestly following the evidence is a form of respect for nature, and honestly communicating one's findings is a form of respect for others. Recognizing the fallibility of both experiment and reason brings humility, and also charity. Knowing that ultimately science only progresses in collaboration builds community. Acknowledging the ways that science and scientists serially fail to embody these virtues does nothing to lessen their importance or authority. His advice in the face of failure is simple: Keep trying.

16

Review:
Shane O'Mara's
Why Torture Doesn't Work: The Neuroscience of Interrogation

By *Moheb Costandi, M.Sc.*

Moheb Costandi, M.Sc., is a neuroscientist-turned-free-lance writer based in London. His work has been published in *Nature, New Scientist, Science*, and *Scientific American*, among others, and he is also author of the long-standing *Neurophilosophy* blog, hosted by *The Guardian*. Costandi has written extensively about neuroethics for the Dana Foundation and serves on the board of directors of the International Neuro-ethics Society. His first book, *50 Human Brain Ideas You Really Need to Know*, was published in 2013 by Quercus, and his second, *Neuroplasticity*, has just been published by MIT Press.

Waterboarding, sleep deprivation, and solitary confinement were some of the tactics outlined and authorized in a series of Bush administration secret legal documents known as the "torture memos," which were made public in 2009. Shane O'Mara's new book casts morality aside to examine whether torture produces reliable information. He reviews existing research in psychology and neuroscience to highlight the impact of torture methods on brain function.

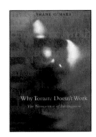

ON APRIL 16, 2009, President Obama released four top secret memoranda, written by top White House's lawyers of the second Bush administration, that came to be known as the "torture memos." These documents not only detailed the "enhanced interrogation" techniques used by the Central Intelligence Agency (CIA) on suspected terrorists detained at Guantánamo Bay Naval Base in Cuba and other secret detention centers around the world, but justified them as effective ways of obtaining sensitive information—and approved their use.

The memos describe ten techniques that were employed with at least 14 suspects, and possibly many, many more: attention grasp, facial hold, facial slap, walling, wall standing, stress positions, cramped confinement, sleep deprivation, insect placed in a confinement box, and waterboarding. Related documents disclose the use of additional methods: One captive was "interrogated for approximately 20 hours a day for seven weeks; given strip searches, sometimes in the presence of female interrogators; forced to wear women's underwear; forcibly injected with large quantities of IV fluid and forced to urinate on himself; led around on a leash; made to bark like a dog; and subjected to cold temperatures"; another was given a lunch of "hummus, pasta with sauce, nuts, and raisins," with the ingredients being "pureed and rectally infused."

The assumption, based on intuition and folk psychology, is that such methods will "break" the captives and enhance their ability to recall incriminating facts. In his book, Shane O'Mara, a professor of experimental brain research at Trinity College, Dublin, casts a highly critical eye over claims by proponents of torture that the CIA's enhanced interrogation techniques can indeed effectively elicit sensitive information.

It would, of course, be highly unethical to mimic the effects of torture under experimental conditions in order to investigate how it affects victims, so there are precious few studies that explicitly aim to do so. O'Mara scours the biomedical literature and describes those human studies that have been performed, along with a wealth of animal experiments that demonstrate the neurological and psychological effects of torture.

Together, this research shows that the effects of torture may in fact be exactly the opposite of what they are intended and claimed to be. It is, for example, well known that memory is reconstructive, rather than reproductive, in nature; our recollection of life events fits newly acquired information into a framework of prior knowledge, expectations, and biases. At best, our memories are not entirely accurate; in extreme cases, they can be misleading or even totally false.

Rather than facilitating memory recall, the various "stressors" experienced under interrogation—the physiological changes that occur in response to the uncomfortable positions in which captives are held, the physical pain inflicted upon them, and the prolonged periods of sleep deprivation to which they are subjected—not only make our recollections less accurate, but also make us more susceptible to confabulating entirely false ones.

Thus, while torture does make captives more likely to confess, the information obtained is not likely to be accurate, and could be pure fabrication—a "broken" captive will almost certainly be extremely confused, to the point where he may be unable to distinguish fact from fantasy, and could very well tell his interrogators what they want to hear in the hope that they will stop torturing him.

While morally opposed to torture, O'Mara sets ethics and values aside to focus on the scientific evidence. He skillfully dissects the claims put forward in the torture memos and systematically demolishes them. In so doing, he presents an airtight argument against the use of enhanced interrogation techniques, on the grounds that they simply do not produce the desired effects.

Unfortunately, though, governments rarely let scientific evidence stand in the way of their political aims, and, as O'Mara makes clear, the second Bush administration endorsed the techniques with little or no concern for their adverse effects, and despite existing evidence that they do not work as intended. Furthermore, the torture memos disingenuously claim that they are not only effective, but also that their use led at least some captives to divulge useful information.

Thus, the US government's decision to use torture is not based on sound reasoning or scientific evidence, and the torture memos served merely to provide a legal basis for the use of these abhorrent practices.

Nevertheless, such practices continue to be used widely by democracies around the world, where they have been sanctioned by both government and various members of the medical profession: Last year, while this book was in production, an independent review revealed that the American Psychological Association—the largest professional body of psychologists in the US—was complicit in the CIA's use of these brutal techniques.

O'Mara devotes the concluding chapters of his book to the psychological factors that contribute to compliance and obedience, and the detrimental effects that torture has on those who inflict it—something the torture memos briefly acknowledge but then gloss over. He goes on to suggest less coercive methods that may work better to elicit useful information—approaches that use virtual reality role-playing, for example—but adds that more research will have to be done to determine whether they are indeed effective.

In the past few decades, neuroscientists have made enormous advances in understanding brain and behavior. O'Mara expresses surprise that the field has not brought this new knowledge to bear on these issues, and his hope that his book will galvanize colleagues in neuroscience, psychology, and psychiatry to become involved in them. *Why Torture Doesn't Work* is exceedingly well written and meticulously researched. It is not an easy book to read—because of its subject matter—but it is a hugely important one.

Endnotes

1
The Changing Face of Recreational Drug Use

1. Zawilska JB, Andrzejczak D. Next generation of novel psychoactive substances on the horizon - A complex problem to face. Drug Alcohol Depend. (2015) 157:1-17.
2. Schifano F, Orsolini L, DuccioPapanti G, Corkery JM. Novel psychoactive substances of interest for psychiatry. World Psychiatry. (2015) 14:15-26.
3. Trecki J, Gerona RR, Schwartz MD. Synthetic cannabinoid-related illnesses and deaths. New Engl J Med. (2015) 373:103-107.
4. Brandt SD, King LA, Evans-Brown M. The new drug phenomenon. Drug Test Analysis. (2014) 6:587-597.
5. Erowid Experiences Vault (2015). https://www.erowid.org/experiences/
6. United Nations Office of Drugs and Crime. World Drug Report (2015). http://www.unodc.org/documents/wdr2015/World_Drug_Report_2015.pdf
7. American Association of Poison Control Centers, Bath Salts Alert (2015). http://www.aapcc.org/alerts/bath-salts/
8. Spiller HA, Ryan ML, Weston RG, Jansen J. Clinical experience with and analytical confirmation of "bath salts" and "legal highs" (synthetic cathinones) in the United States. Clin Toxicol. (2011) 49:499-505.
9. Drug Enforcement Administration, Department of Justice. Establishment of drug codes for 26 substances. Final rule. Fed Regist. (2013) 78(3):664-6.
10. http://www.deadiversion.usdoj.gov/fed_regs/rules/2013/fr0412_2.htm
11. Drug Enforcement Administration. Special report: synthetic cannabinoids and cathinones reported in NFLIS, 2010-2013 (2014). http://www.deadiversion.usdoj.gov/nflis/spec_rpt_CathCan_2013.pdf
12. Baumann MH, Partilla JS, Lehner KR. Psychoactive "bath salts": not so soothing. Eur J Pharmacol. (2013) 698:1-5.
13. Baumann MH, Partilla JS, Lehner KR, Thorndike EB, Hoffman AF, Holy M, Rothman RB, Goldberg SR, Lupica CR, Sitte HH, Brandt SD, Tella SR, Cozzi NV, Schindler CW. Powerful cocaine-like actions of 3,4-methylenedioxypyrovalerone (MDPV), a principal constituent of psychoactive 'bath salts' products. Neuropsychopharmacology. (2013) 38:552-62
14. Simmler LD, Buser TA, Donzelli M, Schramm Y, Dieu LH, Huwyler J, Chaboz S, Hoener MC, Liechti ME. Pharmacological characterization of designer cathinones in vitro. Br J Pharmacol. (2013) 168:458-70.
15. Baumann MH, Ayestas MA Jr, Partilla JS, Sink JR, Shulgin AT, Daley PF, Brandt SD, Rothman RB, Ruoho AE, Cozzi NV. The designer methcathinone analogs,

mephedrone and methylone, are substrates for monoamine transporters in brain. Neuropsychopharmacology. (2012) 37:1192-203.

16. Watterson LR, Olive MF. Synthetic cathinones and their rewarding and reinforcing effects in rodents. Adv Neurosci. (2014) Jun 4:209875.

17. Schindler CW, Thorndike EB, Goldberg SR, Lehner KR, Cozzi NV, Brandt SD, Baumann MH. Reinforcing and neurochemical effects of the "bath salts" constituents 3,4-methylenedioxypyrovalerone (MDPV) and 3,4-methylenedioxy-N-methylcathinone (methylone) in male rats. Psychopharmacology. (2015) in press.

18. American Association of Poison Control Centers. Synthetic Cannabinoids Alert (2015). http://www.aapcc.org/alerts/synthetic-cannabinoids/

19. Gunderson EW, Haughey HM, Ait-Daoud N, Joshi AS, Hart CL. "Spice" and "K2" herbal highs: a case series and systematic review of the clinical effects and biopsychosocial implications of synthetic cannabinoid use in humans. Am J Addict. (2012) 21:320-356.

20. Wiley JL, Marusich JA, Huffman JW. Moving around the molecule: relationship between chemical structure and in vivo activity of synthetic cannabinoids. Life Sci. (2014) 97:55-63.

21. Wiley JL, Marusich JA, Martin BR, Huffman JW. 1-Pentyl-3-phenylacetylindoles and JWH-018 share in vivo cannabinoid profiles in mice. Drug Alcohol Depend. (2012) 123:148-153.

22. Wiley JL, Marusich JA, Lefever TW, Grabenauer M, Moore KN, Thomas BF. Cannabinoids in disguise: 9-tetrahydrocannabinol-like effects of tetramethylcyclopropyl ketone indoles. Neuropharmacology. (2013) 75:145-54.

23. Pertwee RG. Ligands that target cannabinoid receptors in the brain: from THC to anandamide and beyond. Addict Biol. (2008) 13:147-59.

24. Hoffman AF, Lupica CR. Synaptic targets of 9-tetrahydrocannabinol in the central nervous system. Cold Spring Harb Perspect Med. (2013) 3: a012237.

25. Fantegrossi WE, Moran JH, Radominska-Pandya A, Prather PL. Distinct pharmacology and metabolism of K2 synthetic cannabinoids compared to (9)-THC: mechanism underlying greater toxicity? Life Sci. (2014) 97:45-54.

26. Järbe TU, Gifford RS. "Herbal incense": designer drug blends as cannabimimetics and their assessment by drug discrimination and other in vivo bioassays. Life Sci. (2014) 97:64-71.

27. Seddon T. Drug policy and global regulatory capitalism: the case of new psychoactive substances (NPS). Int J Drug Policy. (2014) 25:1019-1024.

28. http://www.sun-sentinel.com/local/broward/fl-flakka-deaths-20150625-story.html

29. https://www.ecstasydata.org/results.php?start=0&search_field=all&s=ethylone

30. Nutt DJ, King LA, Nichols DE. Effects of Schedule I drug laws on neuroscience research and treatment innovation. Nat Rev Neurosci. (2013) 14:577-585.

31. Musselman ME, Hampton JP. "Not for human consumption": a review of emerging designer drugs. Pharmacotherapy. (2014) 34:745-57.

32. Nelson ME, Bryant SM, Aks SE. Emerging drugs of abuse. Emerg Med Clin North Am. (2014) 32:1-28.

33. Berry-Cabán CS, Kleinschmidt PE, Rao DS, Jenkins J. Synthetic cannabinoid and cathinone use among US soldiers. US Army Med Dep J. (2012) Oct-Dec:19-24.

34. Concheiro M, Anizan S, Ellefsen K, Huestis MA. Simultaneous quantification of 28 synthetic cathinones and metabolites in urine by liquid chromatography-high resolution mass spectrometry. Anal Bioanal Chem. (2013) 405:9437-9448.
35. Scheidweiler KB, Jarvis MJ, Huestis MA. Nontargeted SWATH acquisition for identifying 47 synthetic cannabinoid metabolites in human urine by liquid chromatography-high-resolution tandem mass spectrometry. Anal Bioanal Chem. (2015) 407:883-897.

2
Lithium to the Rescue

1. Manji HK, and Zarate CA. "Molecular and cellular mechanisms underlying mood stabilization in bipolar disorder: implications for the development of improved therapeutics" Molecular Psychiatry 7 Supplement 1 (2002): S1-7.
2. Jope RS, and Bijur GN. "Mood stabilizers, glycogen synthase kinase-3 and cell survival" Molecular Psychiatry 7 Supplement 1 (2002): S35-45.
3. Klein, P. S., and Melton, D. A. "A molecular mechanism for the effect of lithium on development" Proc. Natl. Acad. Sci. USA 93 (1996): 8455-8459.
4. Beurel, E. and Jope, R. S. "The paradoxical pro- and anti-apoptotic actions of GSK3 in the intrinsic and extrinsic apoptosis signaling pathways" Progress in Neurobiology 79 (2006): 173-189
5. Hanson ND, Nemeroff CB, and Owens MJ. "Lithium, but not fluoxetine or the corticotropin-releasing factor receptor 1 receptor antagonist R121919, increases cell proliferation in the adult dentate gyrus" Journal of Pharmacology and Experimental Therapeutics 337 (2011):180-186
6. Dudev T, and Lim C. "Competition between Li+ and Mg2+ in metalloproteins. Implications for lithium therapy" Journal of the American Chemical Society 133 (2011) 9506-9515.
7. Beurel E, Grieco SF, and Jope RS. "Glycogen synthase kinase-3 (GSK3): regulation, actions, and diseases" Pharmacology and Therapeutics 148 (2015): 114-131
8. Martinez A, and Perez DI, Gil C. "Lessons learnt from glycogen synthase kinase 3 inhibitors development for Alzheimer's disease" Current Topics in Medicinal Chemistry 13 (2013): 1808-1819
9. Gerhard T, Devanand DP, Huang C, Crystal S, and Olfson M. "Lithium treatment and risk for dementia in adults with bipolar disorder: population-based cohort study" British Journal of Psychiatry 207 (2015): 46-51
10. Sutherland C, and Duthie AC. "Invited commentary on … Lithium treatment and risk for dementia in adults with bipolar disorder" British Journal of Psychiatry 207 (2015): 52-54
11. Matsunaga S, Kishi T, Annas P, Basun H, Hampel H, and Iwata N. "Lithium as a Treatment for Alzheimer's Disease: A Systematic Review and Meta-Analysis" Journal of Alzheimers Disease 48 (2015): 403-410
12. Morales-García JA, Susín C, Alonso-Gil S, Pérez DI, Palomo V, Pérez C, Conde S, Santos A, Gil C, Martínez A, and Pérez-Castillo A. "Glycogen synthase kinase-3 inhibitors as potent therapeutic agents for the treatment of Parkinson disease" ACS Chemical Neuroscience 4 (2013): 350-360
13. Chuang DM, Wang Z, and Chiu CT. "GSK-3 as a Target for Lithium-Induced Neuroprotection Against Excitotoxicity in Neuronal Cultures and Animal Mod-

els of Ischemic Stroke" Frontiers in Molecular Neuroscience 4 (2011): 15.

14. Leeds PR, Yu F, Wang Z, Chiu CT, Zhang Y, Leng Y, Linares GR, and Chuang DM. "A New Avenue for Lithium: Intervention in Traumatic Brain Injury" ACS Chemical Neuroscience 5 (2014): 422-433

15. Beurel E. "Regulation by glycogen synthase kinase-3 of inflammation and T cells in CNS diseases" Frontiers in Molecular Neuroscience 4 (2011):18.

16. Martin, M., Rehani, K., Jope, R. S., and Michalek, S. M. "Toll-like receptor-mediated cytokine production is differentially regulated by glycogen synthase kinase 3" Nature Immunology 6 (2005): 777-784.

17. Beurel E, Kaidanovich-Beilin O, Yeh WI, Song L, Palomo V, Michalek SM, Woodgett JR, Harrington LE, Eldar-Finkelman H, Martinez A, and Jope RS. "Regulation of Th1 cells and experimental autoimmune encephalomyelitis by glycogen synthase kinase-3" Journal of Immunology 190 (2013): 5000-5011.

18. Mines MA, and Jope RS. "Glycogen synthase kinase-3: a promising therapeutic target for Fragile X syndrome" Frontiers in Molecular Neuroscience 4 (2011): 35.

19. Berry-Kravis E, Sumis A, Hervey C, Nelson M, Porges SW, Weng N, Weiler IJ, and Greenough WT. "Open-label treatment trial of lithium to target the underlying defect in fragile X syndrome" Journal of Dev. Behav. Pediatrics 29 (2008): 293-302.

20. Beurel E, and Jope RS. "Inflammation and lithium: clues to mechanisms contributing to suicide-linked traits" Translational Psychiatry 4 (2014): e488.

21. Fels, A. "Should we all take a bit of lithium?" New York Times, (2014) Sept. 14.

22. Vita A, De Peri L, and Sacchetti E. "Lithium in drinking water and suicide prevention: a review of the evidence" International Clinical Psychopharmacology 30 (2015): 1-5.

23. Chiu CT, Wang Z, Hunsberger JG, and Chuang DM. "Therapeutic potential of mood stabilizers lithium and valproic acid: beyond bipolar disorder" Pharmacological Reviews 65 (2013): 105-142.

3
The Malignant Protein Puzzle

1. Bender, H. S., Marshall Graves, J. A. & Deakin, J. E. Pathogenesis and molecular biology of a transmissible tumor in the Tasmanian devil. Annu Rev Anim Biosci 2, 165-187, doi:10.1146/annurev-animal-022513-114204 (2014).

2. Strakova, A. & Murchison, E. P. The cancer which survived: insights from the genome of an 11000 year-old cancer. Curr Opin Genet Dev 30, 49-55, doi:10.1016/j.gde.2015.03.005 (2015).

3. Metzger, M. J., Reinisch, C., Sherry, J. & Goff, S. P. Horizontal transmission of clonal cancer cells causes leukemia in soft-shell clams. Cell 161, 255-263, doi:10.1016/j.cell.2015.02.042 (2015).

4. Schwartz, M. How the Cows Turned Mad. (University of California Press, 2003).

5. Cuille, J. & Chelle, P.-L. La maladie dite tremblante du mouton est-elle inoculable? Comptes Rendus de l'Academie des Sciences 203, 1552-1554 (1936).

6. Gajdusek, D. C. Unconventional viruses and the origin and disappearance of kuru. Science 197, 943-960 (1977).

7. Alper, T., Cramp, W. A., Haig, D. A. & Clarke, M. C. Does the agent of scrapie replicate without nucleic acid? Nature 214, 764-766 (1967).
8. Griffith, J. S. Self-replication and scrapie. Nature 215, 1043-1044 (1967).
9. Prusiner, S. B. Madness and Memory. (Yale University Press, 2014).
10. Couzin-Frankel, J. Scientific community. The prion heretic. Science 332, 1024-1027, doi:10.1126/science.332.6033.1024 (2011).
11. Jucker, M. & Walker, L. C. Self-propagation of pathogenic protein aggregates in neurodegenerative diseases. Nature 501, 45-51, doi:10.1038/nature12481 (2013).
12. Prusiner, S. B. Biology and genetics of prions causing neurodegeneration. Annu Rev Genet 47, 601-623, doi:10.1146/annurev-genet-110711-155524 (2013).
13. Walker, L. C. & Jucker, M. Seeds of dementia. Sci Am 308, 52-57 (2013).
14. Walker, L. C. & Jucker, M. Neurodegenerative diseases: expanding the prion concept. Annu Rev Neurosci 38, 87-103, doi:10.1146/annurev-neuro-071714-033828 (2015).
15. Lansbury, P. T., Jr. & Caughey, B. The chemistry of scrapie infection: implications of the 'ice 9' metaphor. Chem Biol 2, 1-5 (1995).
16. Imran, M. & Mahmood, S. An overview of human prion diseases. Virology journal 8, 559, doi:10.1186/1743-422X-8-559 (2011).
17. Imran, M. & Mahmood, S. An overview of animal prion diseases. Virology journal 8, 493, doi:10.1186/1743-422X-8-493 (2011).
18. DeArmond, S. J., Ironside, J. W., Bouzamondo-Bernstein, E., Peretz, D. & Fraser, J. R. in Prion Biology and Diseases (ed S. B. Prusiner) 777-856 (Cold Spring Harbor Laboratory Press, 2004).
19. Hardy, J. & Selkoe, D. J. The amyloid hypothesis of Alzheimer's disease: progress and problems on the road to therapeutics. Science 297, 353-356, doi:10.1126/science.1072994 297/5580/353 [pii] (2002).
20. Goudsmit, J. et al. Evidence for and against the transmissibility of Alzheimer disease. Neurology 30, 945-950 (1980).
21. Baker, H. F., Ridley, R. M., Duchen, L. W., Crow, T. J. & Bruton, C. J. Induction of beta (A4)-amyloid in primates by injection of Alzheimer's disease brain homogenate. Comparison with transmission of spongiform encephalopathy. Mol Neurobiol 8, 25-39 (1994).
22. Brettschneider, J., Del Tredici, K., Lee, V. M. & Trojanowski, J. Q. Spreading of pathology in neurodegenerative diseases: a focus on human studies. Nat Rev Neurosci 16, 109-120, doi:10.1038/nrn3887 (2015).
23. Goedert, M. NEURODEGENERATION. Alzheimer's and Parkinson's diseases: The prion concept in relation to assembled Abeta, tau, and alpha-synuclein. Science 349, 1255555, doi:10.1126/science.1255555 (2015).
24. King, O. D., Gitler, A. D. & Shorter, J. The tip of the iceberg: RNA-binding proteins with prion-like domains in neurodegenerative disease. Brain Res 1462, 61-80, doi:10.1016/j.brainres.2012.01.016 (2012).
25. Ayers, J. I., Fromholt, S. E., O'Neal, V. M., Diamond, J. H. & Borchelt, D. R. Prion-like propagation of mutant SOD1 misfolding and motor neuron disease spread along neuroanatomical pathways. Acta Neuropathol, doi:10.1007/s00401-015-1514-0 (2015).
26. Jaunmuktane, Z. et al. Evidence for human transmission of amyloid-beta pathology and cerebral amyloid angiopathy. Nature 525, 247-250, doi:10.1038/nature15369 (2015).

27. Holt, R. I. & Sonksen, P. H. Growth hormone, IGF-I and insulin and their abuse in sport. Br J Pharmacol 154, 542-556, doi:10.1038/bjp.2008.99 (2008).
28. Frontzek, K., Lutz, M. I., Aguzzi, A., Kovacs, G. G. & Budka, H. Amyloid-beta pathology and cerebral amyloid angiopathy are frequent in iatrogenic Creutzfeldt-Jakob disease after dural grafting. Swiss Med Wkly 146, w14287, doi:10.4414/smw.2016.14287 (2016).
29. Wickner, R. B. et al. Yeast prions: structure, biology, and prion-handling systems. Microbiol Mol Biol Rev 79, 1-17, doi:10.1128/MMBR.00041-14 (2015).
30. Maji, S. K. et al. Functional amyloids as natural storage of peptide hormones in pituitary secretory granules. Science 325, 328-332, doi:10.1126/science.1173155 (2009).
31. Fioriti, L. et al. The Persistence of Hippocampal-Based Memory Requires Protein Synthesis Mediated by the Prion-like Protein CPEB3. Neuron 86, 1433-1448, doi:10.1016/j.neuron.2015.05.021 (2015).
32. Haley, N. J. & Hoover, E. A. Chronic wasting disease of cervids: current knowledge and future perspectives. Annu Rev Anim Biosci 3, 305-325, doi:10.1146/annurev-animal-022114-111001 (2015).

4
Imaging the Neural Symphony

1. Masters, B. R. and P. T. So (2004). "Antecedents of two-photon excitation laser scanning microscopy." Microsc Res Tech 63(1): 3-11.
2. Denk, W., J. H. Strickler and W. W. Webb (1990). "Two-photon laser scanning microscopy." Science 248: 73-76.
3. Denk, W. and K. Svoboda (1997). "Photon upmanship: why multiphoton imaging is more than a gimmick." Neuron 18: 351-357.
4. Tsien, R. Y. (1998). "The green fluorescent protein." Annu Rev Biochem 67: 509-544.
5. Trachtenberg, J. T., B. E. Chen, G. W. Knott, G. Feng, J. R. Sanes, E. Welker and K. Svoboda (2002). "Long-term in vivo imaging of experience-dependent synaptic plasticity in adult cortex." Nature 420(6917): 788-794.
6. Holtmaat, A. and K. Svoboda (2009). "Experience-dependent structural synaptic plasticity in the mammalian brain." Nat Rev Neurosci 10(9): 647-658.
7. Chen, T. W., T. J. Wardill, Y. Sun, S. R. Pulver, S. L. Renninger, A. Baohan, E. R. Schreiter, R. A. Kerr, M. B. Orger, V. Jayaraman, L. L. Looger, K. Svoboda and D. S. Kim (2013). "Ultrasensitive fluorescent proteins for imaging neuronal activity." Nature 499(7458): 295-300.
8. Peron, S. P., J. Freeman, V. Iyer, C. Guo and K. Svoboda (2015). "A Cellular Resolution Map of Barrel Cortex Activity during Tactile Behavior." Neuron 86(3): 783-799.
9. Zeisel, A., A. B. Munoz-Manchado, S. Codeluppi, P. Lonnerberg, G. La Manno, A. Jureus, S. Marques, H. Munguba, L. He, C. Betsholtz, C. Rolny, G. Castelo-Branco, J. Hjerling-Leffler and S. Linnarsson (2015). "Brain structure. Cell types in the mouse cortex and hippocampus revealed by single-cell RNA-seq." Science 347(6226): 1138-1142.
10. Luo, L., E. M. Callaway and K. Svoboda (2008). "Genetic dissection of neural circuits." Neuron 57(5): 634-660.
11. Deisseroth, K. (2011). "Optogenetics." Nature methods 8(1): 26-29.

5
A New Approach for Epilepsy

1. Wilson JV, Reynolds EH. (1990). Texts and documents: translation and analysis of a cuneiform text forming part of a Babylonian treatise on epilepsy. Med. Hist. 34:185–198.
2. Todman D. (2008). Epilepsy in the Graeco-Roman world: Hippocratic medicine and Asklepian temple medicine compared. J. Hist. Neurosci. 17:435–441.
3. Sada N, Lee S, Katsu T, Otsuki T, Inoue T. (2015) Epilepsy treatment. Targeting LDH enzymes with a stiripentol analog to treat epilepsy. Science. 347:1362-1367.
4. Klepper J, Leiendecker B, Bredahl R, Athanassopoulos S, Heinen F, Gertsen E, Flörcken A, Metz A, Voit T. (2002) Introduction of a ketogenic diet in young infants. J Inherit Metab Dis. 25:449-460.
5. Muzykewicz DA, Lyczkowski DA, Memon N, Conant KD, Pfeifer HH, Thiele EA. (2009) Efficacy, safety, and tolerability of the low glycemic index treatment in pediatric epilepsy. Epilepsia. 50:1118-1126.
6. Wu YJ, Zhang LM, Chai YM, Wang J, Yu LF, Li WH, Zhou YF, Zhou SZ. (2016) Six-month efficacy of the Ketogenic diet is predicted after 3 months and is unrelated to clinical variables. Epilepsy Behav. 55:165-9.
7. Mächler P, Wyss MT, Elsayed M, Stobart J, Gutierrez R, von Faber-Castell A, Kaelin V, Zuend M, San Martín A, Romero-Gómez I, Baeza-Lehnert F, Lengacher S, Schneider BL, Aebischer P, Magistretti PJ, Barros LF, Weber B. (2016) In Vivo Evidence for a Lactate Gradient from Astrocytes to Neurons. Cell Metab. 23:94-102.
8. Tsacopoulos M, Magistretti PJ. (1996) Metabolic coupling between glia and neurons. J. Neurosci. 16:877-885.
9. Bélanger M, Allaman I, Magistretti PJ. (2011) Brain energy metabolism: focus on astrocyte-neuron metabolic cooperation. Cell Metab. 14:724-738.
10. Gonzalez SV, Nguyen NH, Rise F, Hassel B. (2005) Brain metabolism of exogenous pyruvate. J Neurochem. 95:284-93.
11. Huttenlocher PR. (1976) Ketonemia and seizures: metabolic and anticonvulsant effects of two ketogenic diets in childhood epilepsy. Pediatr Res. 10:536-540.
12. Lynch BA, Lambeng N, Nocka K, Kensel-Hammes P, Bajjalieh SM, Matagne A, Fuks B. (2004) The synaptic vesicle protein SV2A is the binding site for the antiepileptic drug levetiracetam. Proc Nat Acad Sci USA 101:9861-9866.
13. Garriga-Canut M, Schoenike B, Qazi R, Bergendahl K, Daley TJ, Pfender RM, Morrison JF, Ockuly J, Stafstrom C, Sutula T, Roopra A. (2006) 2-Deoxy-D-glucose reduces epilepsy progression by NRSF-CtBP-dependent metabolic regulation of chromatin structure. Nat. Neurosci. 9:1382–1387.

7
Drinking Water and the Developing Brain

1. Attina, T. M., & Trasande, L. (2013). Economic costs of childhood lead exposure in low- and middle-income countries. Environmental Health Perspectives, 121(9), 1097-1102. doi:ehp.1206424 [pii]
2. Boutwell, B.B., Nelson E.J., Emo B., Vaughn, M> G., Schoofman, M., Rosenfeld, R. et al. (2016). The intersection of aggregate level lead exposure and crime.

Environmental Research 148, 79-85.

3. Trasande, L., Schechter, C., Haynes, K. A., & Landrigan, P. J. (2006). Applying cost analyses to drive policy that protects children: Mercury as a case study. Annals of the New York Academy of Sciences, 1076, 911-923. doi:1076/1/911 [pii]

4. Trasande, L., Zoeller, R. T., Hass, U., Kortenkamp, A., Grandjean, P., Myers, J. P., . . . Heindel, J. J. (2016). Burden of disease and costs of exposure to endocrine disrupting chemicals in the european union: An updated analysis. Andrology, doi:10.1111/andr.12178 [doi]

5. Wright, J. P., Dietrich, K. N., Ris, M. D., Hornung, R. W., Wessel, S. D., Lanphear, B. P., . . . Rae, M. N. (2008). Association of prenatal and childhood blood lead concentrations with criminal arrests in early adulthood. PLoS Medicine, 5(5), e101. doi:10.1371/journal.pmed.0050101 [doi]

6. Needleman, H. L. (1990). What can the study of lead teach us about other toxicants? Environmental Health Perspectives, 86, 183-189.

7. Budday, S., Steinmann, P., & Kuhl, E. (2015). Physical biology of human brain development. Frontiers in Cellular Neuroscience, 9, 10.3389/fncel.2015.00257. doi:10.3389/fncel.2015.00257 [doi]

8. Harrill, J. A., Chen, H., Streifel, K. M., Yang, D., Mundy, W. R., & Lein, P. J. (2015). Ontogeny of biochemical, morphological and functional parameters of synaptogenesis in primary cultures of rat hippocampal and cortical neurons. Molecular Brain, 8, 10.1186/s13041-015-0099-9. doi:99 [pii]

9. Gioiosa, L., Parmigiani, S., Vom Saal, F. S., & Palanza, P. (2013). The effects of bisphenol A on emotional behavior depend upon the timing of exposure, age and gender in mice. Hormones and Behavior, 63(4), 598-605. doi:10.1016/j.yhbeh.2013.02.016 [doi]

10. Sanders, A., Henn, B. C., & Wright, R. O. (2015). Perinatal and childhood exposure to cadmium, manganese and metal mixtures and effects on cognition and behavior: A review of recent literature. Current Environmental Health Reports, 2, 284.

11. Frazier, T. W., Thompson, L., Youngstrom, E. A., Law, P., Hardan, A. Y., Eng, C., & Morris, N. (2014). A twin study of heritable and shared environmental contributions to autism (44(8) ed.) J Autism Dev Disord. doi:- 10.1007/s10803-014-2081-2

12. Keil, K. P., & Lein, P. J. (2016). DNA methylation: A mechanism linking environmental chemical exposures to risk of autism spectrum disorders? Environmental Epigenetics, 2(1), dvv012. Epub 2016 Jan 30 doi:10.1093/eep/dvv012. doi:10.1093/eep/dvv012 [doi]

13. Newschaffer, C. J., Fallin, D., & Lee, N. L. (2002). Heritable and nonheritable risk factors for autism spectrum disorders. Epidemiologic Reviews, 24(2), 137-153. doi:10.1093/epirev/mxf010

14. Rossignol, D. A., Genuis, S. J., & Frye, R. E. (2014). Environmental toxicants and autism spectrum disorders: A systematic review. Translational Psychiatry, 4, e360. doi:10.1038/tp.2014.4 [doi]

15. Maccani, J. Z., Koestler, D. C., Houseman, E. A., Armstrong, D. A., Marsit, C. J., & Kelsey, K. T. (2015). DNA methylation changes in the placenta are associated with fetal manganese exposure. Reproductive Toxicology (Elmsford, N.Y.), 57, 43-49. doi:10.1016/j.reprotox.2015.05.002 [doi]

16. Balazs, C., Morello-Frosch, R., Hubbard, A., & Ray, I. (2011). Social disparities in

nitrate-contaminated drinking water in California's San Joaquin valley. Environmental Health Perspectives, 119(9), 1272-1278. doi:10.1289/ehp.1002878 [doi]

17. Cushing, L., Morello-Frosch, R., Wanter, M., & Paston, M. (2015). The haves, the have-nots, and the health of everyone: The relationship between social inequality and environmental quality (36: 109-209 ed.) - Annual Reviews. doi:- 10.1146/annurev-publhealth-031914-122646

18. Waldron, H. A. (1973). Lead poisoning in the ancient world. Medical History, 17.04, 391 - 399.

19. Beattie AD, Moore MR, Goldberg, Finlayson MJ, Graham JF, Mackie EM, Main JC, McLaren DA, Murdoch KM, Steward GT(1975). Role of chronic low-level lead exposure in the aetiology of mental retardation. Lancet. 1(7907):589-92

20. Karagas MR, Gossai A, Pierce B, Ahsan H.(2015). Drinking Water Arsenic Contamination, Skin Lesions, and Malignancies: A Systematic Review of the Global Evidence. Curr Environ Health Rep. (1):52-68. doi: 10.1007/s40572-014-0040-x.

21. Shibata T, Meng C, Umoren J, West H. Risk Assessment of Arsenic in Rice Cereal and Other Dietary Sources for Infants and Toddlers in the U.S. (2016). Int J Environ Res Public Health. r13(4). pii: E361. doi: 10.3390/ijerph13040361.

22. Ema, M., Gamo, M., & Honda, K. (2016). Developmental toxicity of engineered nanomaterials in rodents. Toxicology and Applied Pharmacology, 299, 47-52. doi:10.1016/j.taap.2015.12.015 [doi]

23. Ema, M., Gamo, M., & Honda, K. (2016). Developmental toxicity of engineered nanomaterials in rodents. Toxicology and Applied Pharmacology, 299, 47-52. doi:10.1016/j.taap.2015.12.015 [doi]

24. Ballatori, N. (2002). Transport of toxic metals by molecular mimicry. Environmental Health Perspectives, 110(Suppl 5), 689-694.

25. Gulson, B. (2008). Stable lead isotopes in environmental health with emphasis on human investigations. Science of the Total Environment, 400(1–3), 75-92. doi:http://dx.doi.org/10.1016/j.scitotenv.2008.06.059

26. Mori, C., Kakuta, K., Matsuno, Y., Todaka, E., Watanabe, M., Hanazato, M., . . . Fukata, H. (2014). Polychlorinated biphenyl levels in the blood of japanese individuals ranging from infants to over 80Å years of age. Environmental Science and Pollution Research International, 21(10), 6434-6439. doi:1965 [pii]

27. Eto, K. (1997). Pathology of minamata disease. Toxicologic Pathology, 25(6), 614-623.

28. Sass, J. B., Haselow, D. T., & Silbergeld, E. K. (2001). Methylmercury-induced decrement in neuronal migration may involve cytokine-dependent mechanisms: A novel method to assess neuronal movement in vitro. Toxicological Sciences 63(1), 74-81.

29. Kwakye, G. F., Paoliello, M. M., Mukhopadhyay, S., Bowman, A. B., & Aschner, M. (2015). Manganese-induced parkinsonism and parkinson's disease: Shared and distinguishable features. International Journal of Environmental Research and Public Health, 12(7), 7519-7540. doi:10.3390/ijerph120707519 [doi]

30. Zoni, S., & Lucchini, R. G. (2013). Manganese exposure: Cognitive, motor and behavioral effects on children: A review of recent findings. Current Opinion in Pediatrics, 25(2), 255-260. doi:10.1097/MOP.0b013e32835e906b [doi]

31. Chung, S. E., Cheong, H. K., Ha, E. H., Kim, B. N., Ha, M., Kim, Y., . . . Oh, S. Y. (2015). Maternal blood manganese and early neurodevelopment: The moth-

ers and childrens environmental health (MOCEH) study. Environmental Health Perspectives, 123(7), 717-722. doi:ehp.1307865 [pii]

32. Rodrigues, E. G., Bellinger, D. C., Valeri, L., Hasan, M. O., Quamruzzaman, Q., Golam, M., . . . Mazumdar, M. (2016). Neurodevelopmental outcomes among 2- to 3-year-old children in bangladesh with elevated blood lead and exposure to arsenic and manganese in drinking water. Environmental Health 15, 44-016-0127-y. doi:10.1186/s12940-016-0127-y [doi]

33. Moser, V. C., Phillips, P. M., Levine, A. B., McDaniel, K. L., Sills, R. C., Jortner, B. S., & Butt, M. T. (2004). Neurotoxicity produced by dibromoacetic acid in drinking water of rats. Toxicological Sciences, 79(1), 112-122. doi:10.1093/toxsci/kfh081 [doi]

34. Boxall, A. B., Rudd, M. A., Brooks, B. W., Caldwell, D. J., Choi, K., Hickmann, S., . . . Van Der Kraak, G. (2012). Pharmaceuticals and personal care products in the environment: What are the big questions? Environmental Health Perspectives, 120(9), 1221-1229. doi:10.1289/ehp.1104477 [doi]

35. Kaushik, G., Huber, D. P., Aho, K., Finney, B., Bearden, S., Zarbalis, K. S., & Thomas, M. A. (2016). Maternal exposure to carbamazepine at environmental concentrations can cross intestinal and placental barriers. Biochemical and Biophysical Research Communications, 474(2), 291-295. doi:10.1016/j.bbrc.2016.04.088 [doi]

36. Conley, J. M., Evans, N., Mash, H., Rosenblum, L., Schenck, K., Glassmeyer, S., . . . Wilson, V. S. (2016). Comparison of in vitro estrogenic activity and estrogen concentrations in source and treated waters from 25 U.S. drinking water treatment plants. The Science of the Total Environment, doi:S0048-9697(16)30303-5 [pii]

37. Ejaredar, M., Nyanza, E. C., Ten Eycke, K., & Dewey, D. (2015). Phthalate exposure and childrens neurodevelopment: A systematic review. Environmental Research, 142, 51-60. doi:10.1016/j.envres.2015.06.014 [doi]

38. Kassotis, C. D., Iwanowicz, L. R., Akob, D. M., Cozzarelli, I. M., Mumford, A. C., Orem, W. H., & Nagel, S. C. (2016). Endocrine disrupting activities of surface water associated with a West Virginia oil and gas industry wastewater disposal site. The Science of the Total Environment, 557-558, 901-910. doi:10.1016/j.scitotenv.2016.03.113 [doi]

39. Bernal, J, Guadano-Ferraz, A, and Morte, B. (2003) Perspectives in the study of throid hormone acton onf braindevelopment and function. Thyroid 13(11), 1005.

40. Zoeller, T. R. (2010). Environmental chemicals targeting thyroid. Hormones (Athens, Greece), 9(1), 28-40.

41. Lefebvre, K.A.., Frame, E.R., Gulland, F., et al (2012). A novel antibody-based biomarker for chronic algal toxic exposure and sub-acute neurotoxicity. PLoSOne, 7(5), e36213.

42. Cheng, Y. S., Zhou, Y., Irvin, C. M., Pierce, R. H., Naar, J., Backer, L. C., . . . Baden, D. G. (2005). Characterization of marine aerosol for assessment of human exposure to brevetoxins. Environmental Health Perspectives, 113(5), 638-643.

43. Ramsdell, J. S., & Zabka, T. S. (2008). In utero domoic acid toxicity: A fetal basis to adult disease in the california sea lion (zalophus californianus). Marine Drugs, 6(2), 262-290. doi:10.3390/md20080013 [doi]

44. Weirich, C. A., & Miller, T. R. (2014). Freshwater harmful algal blooms: Toxins

and children's health. Current Problems in Pediatric and Adolescent Health Care 44(1), 2.

45. Doucette, T. A., & Tasker, R. A. (2015). Perinatal domoic acid as a neuroteratogen. Current Topics in Behavioral Neurosciences, doi:doi:10.1007/7854_2015_417

46. Alessio, L., Campagna, M., & Lucchini, R. (2007). From lead to manganese through mercury: Mythology, science, and lessons for prevention. American Journal of Industrial Medicine, 50(11), 779-787. doi:10.1002/ajim.20524

47. Fritsche, E., Alm, H., Baumann, J., Geerts, L., Håkansson, H., Masjosthusmann, S., & Witters, H. (2015). Literature review on in vitro and alternative developmental neurotoxicity (DNT) testing methods1. European Food Safety Authority, EN-778

48. Goudarzi, H., Nakajima, S., Ikeno, T., Sasaki, S., Kobayashi, S., Miyashita, C., . . . Kishi, R. (2016). Prenatal exposure to perfluorinated chemicals and neurodevelopment in early infancy: The Hokkaido study. The Science of the Total Environment, 541, 1002-1010. doi:10.1016/j.scitotenv.2015.10.017 [doi]

8
Making Mental Health a Global Priority

1. Marquez, P.V. (2015). "Is Unemployment Bad for Your Health? The World Bank Blogs, January 22, 2015. Available at http://blogs.worldbank.org/health/unemployment-bad-your-health.

2. Marquez, P.V. (2015). "Is Violence a Public Health Problem?" The World Bank Blogs, April 8, 2015. Available at http://blogs.worldbank.org/health/violence-public-health-problem.

3. United Nations, World Bank Group, European Union and African Development Bank. (2015). Recovering from the Ebola Crisis. New York: UNDP. Available at: http://www.undp.org/content/undp/en/home/librarypage/crisis-prevention-and-recovery/recovering-from-the-ebola-crisis---full-report.html.

4. Government of Nepal. (2015). Nepal Earthquake 2015. Post Disaster Needs Assessment. Vol. A: Key Findings. Singha Durbar, Kathmandu: National Planning Commission.

5. Some of below data and information drawn from: Mnookin S., et al. Out of the shadows: making mental health a global development priority report. Geneva: World Bank Group and World Health Organization, 2016. Available at: http://documents.worldbank.org/curated/en/270131468187759113/Out-of-the-shadows-making-mental-health-a-global-development-priority.

6. Patel, V. and S. Saxena (2014). "Transforming lives, enhancing communities — innovations in global mental health." New England Journal of Medicine 370, no. 6:498-501; Helliwell, J.F., et al. (2013). World Happiness Report. Available at http://unsdsn.org/wp-content/uploads/2014/02/WorldHappinessReport2013_online.pdf.

7. World Health Organization (n.d.). "10 Facts on Mental Health." Available at http://www.who.int/features/factfiles/mental_health/mental_health_facts/en/; De Silva, M. and J. Roland, on behalf of the Global Health and Mental Health All-Party Parliamentary Groups (2014). Mental health for sustainable development. London, UK.

8. WHO (n.d.) "Health Statistics and Information Systems: Estimates for 2000–

2012." Available at http://www.who.int/healthinfo/global_burden_disease/esti-mates/en/index2.html.

9. Bloom DE, Cafi ero E, Jané-Llopis E, et al. The global economic burden of non-communicable diseases. Geneva: World Economic Forum, 2011.

10. Hewlett, E. and Moran, V. (2014). Making Mental Health Count: The Social and Economic Costs of Neglecting Mental Health Care. OECD Health Policy Studies, OECD Publishing.

11. Marquez, P.M. (2014). "Mental Health: Time for a Broader Agenda".? The World Bank Blogs, April 22, 2014. Available at http://blogs.worldbank.org/health/mental-health-time-broader-agenda.

12. World Bank Live. Out of the Shadows: Making Mental Health a Global Development Priority. April 13–14, 2016. http://live.worldbank.org/out-ofthe-shad-ows-making-mental-health-a-global-development-priority.

13. WHO. (2013). Mental Health Action Plan 2013-2020. WHO, Geneva.

14. World Bank Group, WHO. Report of Proceedings of Event "Out of the Shad-ows: Making Mental Health a Global Development Priority." Washington, DC: World Bank Group, 2016. Available at: http://www.worldbank.org/en/topic/health/brief/mental-health.

15. Kleinman, A., Lockwood Estrin, G., Usmani, S., Chisholm, D., Marquez, P.V., Evans, T.G., and Saxena, S. Time for mental health to come out of the shadows. Lancet 2016: 387: 2274-2275. Available at: http://www.thelancet.com/jour-nals/lancet/article/PIIS0140-6736(16)30655-9/abstract.

16. WHO (2015). Mental Health Atlas-2014, WHO, Geneva.

17. Chisholm D, Sweeny K, Sheehan P, et al. Scaling-up treatment of depression and anxiety: a global return on investment analysis. Lancet Psychiatry 2016; 3: 415–24.

18. Whiteford, H., et al. (2013). "Global burden of disease attributable to mental and substance use disorders: findings from the Global Burden of Disease Study 2010." Lancet 382: 1575-86.

19. Marquez, P.V., and Saxena, S. (2016). "Mental Health Parity in the Glob-al Health and Development Agenda". The World Bank Blogs, April 4, 2016. Available at http://blogs.worldbank.org/health/mental-health-parity-glob-al-health-and-development-agenda.

20. Kennedy, P., and Fried, S. (2015). A Common Struggle: A Personal Journey Through the Past and Future of Mental Illness and Addiction. Blue Rider Press.

21. See Mnookin S., et al. (2016) above, and Sorel, E.R., et al. (2005) "Populations' Mental Health in Post Conflict Contexts." Advances in Psychiatry Second Vol-ume (2005): 163.

22. See above Mnookin S., et al.(2016), and Bolton, P., et al. (2002): "Prevalence of depression in rural Rwanda based on symptom and functional criteria." The Journal of Nervous and Mental Disease 190, no. 9:631-637; Chung, R. and M. Kagawa-Singer (1993). "Predictors of psychological distress among Southeast Asian refugees." Social Science & Medicine 36, no. 5: 631-639; De Jong, J., et al. (2001). "Lifetime events and posttraumatic stress disorder in 4 post conflict settings." JAMA 286, no. 5:555-562; De Jong, J., et al. (2003) "Common mental disorders in postconflict settings." Lancet 361, no. 9375:2128-2130; Dubois, V., et al. (2004). "Household survey of psychiatric morbidity in Cambodia." Interna-tional Journal of Social Psychiatry 50, no. 2: 174-185; Karam, E.G., et al. (2008).

"Lifetime prevalence of mental disorders in Lebanon: first onset, treatment, and exposure to war." PLOS Medicine 5, no. 4:e61; Mollica, R.F., et al. (1997) "Effects of war trauma on Cambodian refugee adolescents' functional health and mental health status." Journal of the American Academy of Child & Adolescent Psychiatry 36, no. 8:1098-1106; Mollica, R.F., et al. (2004). "Mental health in complex emergencies." Lancet 364, no. 9450:2058-2067; Pham, P.N, et al. (2004). "Trauma and PTSD symptoms in Rwanda: implications for attitudes toward justice and reconciliation." JAMA 292, no. 5: 602-612; Silove, D., et al. (2008) "Estimating clinically relevant mental disorders in a rural and an urban setting in post-conflict Timor Leste." Archives of General Psychiatry 65, no. 10:1205-1212.

23. Mollica, R.F., et al. (2014) "The enduring mental health impact of mass violence: A community comparison study of Cambodian civilians living in Cambodia and Thailand." International Journal of Social Psychiatry 60, no. 1:6-20.

24. Kim, I. (2015). "Beyond Trauma: Post-resettlement Factors and Mental Health Outcomes Among Latino and Asian Refugees in the United States." Journal of Immigrant and Minority Health:1-9.

25. WHO. (2013) Building Back Better: Sustainable mental health care after emergencies, WHO, Geneva.

26. See Rise Asset Development: http://www.riseassetdevelopment.com/.

27. Farrington, C., et al. (2014). "mHealth and global mental health: still waiting for the mH2 wedding?" Globalization and Health 10:17; Renton, T., et al. (2014). "Web-Based Intervention Programs for Depression: A Scoping Review and Evaluation." Journal of Medical Internet Research 16, no.9:e209; Sarasohn- Kahn, J. (2012). "The Online Couch: Mental Health Care on the Web." California Healthcare Foundation. Available at http://www.chcf.org/~/media/MEDIA%20LIBRARY%20Files/PDF/O/PDF%20OnlineCouchMentalHealthWeb.pdf.

28. Aboujaoude, E. et al. (2015). "Telemental health: A status update." World Psychiatry 14, no. 2:223-30; BinDhim, N.F., et al. (2015) "Depression screening via a smartphone app: cross-country user characteristics and feasibility." Journal of the American Medical Informatics Association 22, no. 1:29-34; Agyapong, V., et al. (2012)." Supportive text messaging for depression and comorbid alcohol use disorder: singleblind randomised trial." Journal of Affective Disorders 141, no. 2:168-176; Kauer, S.D., et al. (2012). "Self-monitoring using mobile phones in the early stages of adolescent depression: randomized controlled trial." Journal of Medical Internet Research 14(3):e67; Reid, S.C., et al (2009) "A mobile phone program to track young people's experiences of mood, stress and coping." Social Psychiatry and Psychiatric Epidemiology 44, no. 6: 501-507.

29. Marquez, P.M. (2013). "Healthier Workplaces = Healthy Profits." ? The World Bank Blogs, January 22, 2013. Available at http://blogs.worldbank.org/health/healthier-workplaces-healthy-profits.

30. Marquez, P.M. (2016). "Time to put "health" into universal health coverage". ? The World Bank Blogs, January 14, 2016. Available at http://blogs.worldbank.org/health/time-put-health-universal-health-coverage.

31. World Economic Forum's Global Agenda Council on Mental Health. Seven actions towards a mentally healthy organisation. http://www.joinmq.org/pages/seven-actions-towards-a-mentally-healthy-organisation.

32. World Bank Group, WHO. Report of Proceedings of Event "Out of the Shad-

ows: Making Mental Health a Global Development Priority." Washington, DC: World Bank Group, 2016. http://www.worldbank.org/en/topic/health/brief/mental-health.

33. See World Bank Live: Opening high level panel of "Out of the Shadows: Making Mental Health a Global Development Priority" event. Washington, DC: World Bank Group, April 13, 2016.   http://live.worldbank.org/out-of-the-shadows-making-mental-health-a-global-development-priority.

34. United Nations. Financing for Development. (Addis Ababa Action Agenda). Outcome document adopted at the Third International Conference on Financing for Development, Addis Ababa, Ethiopia, 13-16 July, 2015. Available at: http://www.un.org/esa/ffd/wp-ontent/uploads/2015/08/AAAA_Outcome.pdf.

35. Marquez, P.V. (2016). "Economic slowdown and financial shocks: can tobacco tax increases help? The World Bank Blogs, Feb 8, 2016. http://blogs.worldbank.org/voices/economic-slowdown-and-financial-shocks-can-tobacco-tax increases-help; Marquez, P.V. (2015). "Making the public health case for tobacco taxation." The World Bank Blogs, July 7, 2015. Available at: http://blogs.worldbank.org/health/making-public-health-case-tobacco-taxation.

36. Akerlof, G.A., Shiller, R.J. (2015). Phishing for Phools: The Economics of Manipulation and Deception. New Jersey: Princeton University Press.

37. Cook, P.J. (2016). Tobacco and Alcohol Taxes: The importance of evidence on public health effects. World Bank Conference: "Winning the Tax Wars: Global Solutions for Developing Countries" Held on May 24, 2016. Available at: http://pubdocs.worldbank.org/en/821891464895789490/PCook-remarks.pdf.

38. Colchero, M.A., Popkin, B.M., Rivera, J.A., Ng, S.W. (2016). "Beverage purchases from stores in Mexico under the excise tax on sugar sweetened beverages: observational study". BMJ 2016;352:h6704.

39. Furman, J. (2016). Six Lessons from the U.S. Experience with Tobacco Taxes. World Bank Conference: "Winning the Tax Wars: Global Solutions for Developing Countries" Held on May 24, 2016. Available at: https://www.whitehouse.gov/sites/default/files/page/files/20160524_cea_tobacco_tax_speech.pdf; Marquez, P.V. (2015). "Running away from "Tobacco Road." The World Bank Blogs, December 12, 2015. Available at: http://blogs.worldbank.org/health/running-away-tobacco-road.

40. Kaiser, K, Bredenkamp, C., Iglesias, R. 2016. Sin Tax Reform in the Philippines: Transforming Public Finance, Health, and Governance for More Inclusive Development. Directions in Development--Countries and Regions;. Washington, DC: World Bank. Available at: https://openknowledge.worldbank.org/handle/10986/24617.

41. Marquez, P.V. (2015). "Shining a light on mental illness: An "invisible disability"? The World Bank Blogs, December 2, 2015. Available at http://blogs.worldbank.org/health/shining-light-mental-illness-invisible-disability; Evans, T, Marquez, P.V., and Saxena, S. (2015). "The zero hour for mental health." The World Bank Blogs, May 4, 2015. Available at http://blogs.worldbank.org/health/zero-hour-mental-health.

9
The Human Conectome Project: Progress and Prospects

1. M.F. Glasser, S.M. Smith, D.S. Marcus, J. Andersson, E.J. Auerbach, T.E.J. Behrens, T.S. Coalson, et al., "The Human Connectome Project's neuroimaging approach," Nature Neuroscience, (in press) (2016).
2. D.C. Van Essen, S.M. Smith, D.M. Barch, T.E. Behrens, E. Yacoub, and K. Ugurbil, "The WU-Minn Human Connectome Project: an overview," Neuroimage, 80 (2013) 62-79.
3. M.F. Glasser, T.S. Coalson, E.C. Robinson, C.D. Hacker, J. Harwell, E. Yacoub, K. Ugurbil, et al., "A multi-modal parcellation of human cerebral cortex," Nature 536 (2016) 171-178 doi 10.1038/nature18933.
4. K. Ugurbil, J. Xu, E.J. Auerbach, S. Moeller, A.T. Vu, J.M. Duarte-Carvajalino, C. Lenglet, et al., "Pushing spatial and temporal resolution for functional and diffusion MRI in the Human Connectome Project," Neuroimage, 80 (2013) 80-104.
5. S.N. Sotiropoulos, S. Jbabdi, J. Xu, J.L. Andersson, S. Moeller, E.J. Auerbach, M.F. Glasser, et al., "Advances in diffusion MRI acquisition and processing in the Human Connectome Project," Neuroimage, 80 (2013) 125-43.
6. J.L. Andersson and S.N. Sotiropoulos, "An integrated approach to correction for off-resonance effects and subject movement in diffusion MR imaging," Neuroimage, 125 (2016) 1063-78.
7. M.F. Glasser, S.N. Sotiropoulos, J.A. Wilson, T.S. Coalson, B. Fischl, J.L. Andersson, J. Xu, et al., "The minimal preprocessing pipelines for the Human Connectome Project," Neuroimage, 80 (2013) 105-24.
8. L. Griffanti, G. Salimi-Khorshidi, C.F. Beckmann, E.J. Auerbach, G. Douaud, C.E. Sexton, E. Zsoldos, et al., "ICA-based artefact removal and accelerated fMRI acquisition for improved resting state network imaging," Neuroimage, 95 (2014) 232-47.
9. G. Salimi-Khorshidi, G. Douaud, C.F. Beckmann, M.F. Glasser, L. Griffanti, and S.M. Smith, "Automatic denoising of functional MRI data: combining independent component analysis and hierarchical fusion of classifiers," Neuroimage, 90 (2014) 449-68.
10. S.M. Smith, D. Vidaurre, C.F. Beckmann, M.F. Glasser, M. Jenkinson, K.L. Miller, T.E. Nichols, et al., "Functional connectomics from resting-state fMRI," Trends Cogn Sci, 17 (2013) 666-82.
11. J.D. Power, A. Mitra, T.O. Laumann, A.Z. Snyder, B.L. Schlaggar, and S.E. Petersen, "Methods to detect, characterize, and remove motion artifact in resting state fMRI," Neuroimage, 84 (2014) 320-41.
12. Z.S. Saad, R.C. Reynolds, H.J. Jo, S.J. Gotts, G. Chen, A. Martin, and R.W. Cox, "Correcting brain-wide correlation differences in resting-state FMRI," Brain Connect, 3 (2013) 339-52.
13. D.C. Van Essen, "Cartography and connectomes," Neuron, 80 (2013) 775-790.
14. D.C. Van Essen and J.H. Maunsell, "Two-dimensional maps of the cerebral cortex," J Comp Neurol, 191 (1980) 255-81.
15. B. Fischl, A. Liu, and A.M. Dale, "Automated manifold surgery: constructing geometrically accurate and topologically correct models of the human cerebral cortex," IEEE Trans Med Imaging, 20 (2001) 70-80.
16. M.F. Glasser and D.C. Van Essen, "Mapping human cortical areas in vivo based

on myelin content as revealed by T1- and T2-weighted MRI," J Neurosci, 31 (2011) 11597-616.

17. M.F. Glasser, M.S. Goyal, T.M. Preuss, M.E. Raichle, and D.C. Van Essen, "Trends and properties of human cerebral cortex: correlations with cortical myelin content," Neuroimage, 93 Pt 2 (2014) 165-75.

18. D.C. Van Essen, H.A. Drury, S. Joshi, and M.I. Miller, "Functional and structural mapping of human cerebral cortex: solutions are in the surfaces," Proc Natl Acad Sci U S A, 95 (1998) 788-95.

19. B. Fischl, M.I. Sereno, R.B. Tootell, and A.M. Dale, "High-resolution intersubject averaging and a coordinate system for the cortical surface," Hum Brain Mapp, 8 (1999) 272-84.

20. H.A. Drury, D.C. Van Essen, C.H. Anderson, C.W. Lee, T.A. Coogan, and J.W. Lewis, "Computerized mappings of the cerebral cortex: a multiresolution flattening method and a surface-based coordinate system," J Cogn Neurosci, 8 (1996) 1-28.

21. E.C. Robinson, S. Jbabdi, M.F. Glasser, J. Andersson, G.C. Burgess, M.P. Harms, S.M. Smith, et al., "MSM: A new flxible framework for multimodal surface matching," Neuroimage, 100 (2014) 414-426.

22. K. Brodmann, "Vergleichende Lokalisationslehre der Grosshirnrinde. Leipzig:," (1909).

23. R. Nieuwenhuys, "The myeloarchitectonic studies on the human cerebral cortex of the Vogt-Vogt school, and their significance for the interpretation of functional neuroimaging data," Brain Struct Funct, 218 (2013) 303-52.

24. C. Vogt and O. Vogt, "Allgemeinere ergebnisse unswerer hirnforschung," J Psychol Neurol, 25 (1919) 279-468.

25. J. Dickson, H. Drury, and D.C. Van Essen, "'The surface management system' (SuMS) database: a surface-based database to aid cortical surface reconstruction, visualization and analysis," Philos Trans R Soc Lond B Biol Sci, 356 (2001) 1277-92.

26. M.R. Hodge, W. Horton, T. Brown, R. Herrick, T. Olsen, M.E. Hileman, M. McKay, et al., "ConnectomeDB--Sharing human brain connectivity data," Neuroimage, 124 (2016) 1102-7.

27. D.S. Marcus, M.P. Harms, A.Z. Snyder, M. Jenkinson, J.A. Wilson, M.F. Glasser, D.M. Barch, et al., "Human Connectome Project informatics: quality control, database services, and data visualization," Neuroimage, 80 (2013) 202-19.

28. D.S. Marcus, T.R. Olsen, M. Ramaratnam, and R.L. Buckner, "The Extensible Neuroimaging Archive Toolkit: an informatics platform for managing, exploring, and sharing neuroimaging data," Neuroinformatics, 5 (2007) 11-34.

29. D. Van Essen, J. Smith, M. Glasser, J. Elam, C. Donahue, D.L. Dierker, E.K. Reid, et al., "The brain analysis of spatial maps and atlases (BALSA) databas," Neuroimage (in press; 10.1016/j.neuroimage.2016.04.002), (2016).

30. S.M. Smith, T.E. Nichols, D. Vidaurre, A.M. Winkler, T.E. Behrens, M.F. Glasser, K. Ugurbil, et al., "A positive-negative mode of population covariation links brain connectivity, demographics and behavior," Nat Neurosci, 18 (2015) 1565-7.

31. B.T. Yeo, F.M. Krienen, M.W. Chee, and R.L. Buckner, "Estimates of segregation and overlap of functional connectivity networks in the human cerebral cortex," Neuroimage, 88 (2014) 212-27.

32. I. Tavor, O. Parker-Jones, R. Mars, S. Smith, B. TE, and S. Jbabdi, "Task-free MRI predicts individual differences in brain activity during task performance," Science, 352 (2016) 216-220.
33. M. Hawrylycz, J.A. Miller, V. Menon, D. Feng, T. Dolbeare, A.L. Guillozet-Bongaarts, A.G. Jegga, et al., "Canonical genetic signatures of the adult human brain," Nat Neurosci, 18 (2015) 1832-44.
34. M.D. Tisdall, M. Reuter, A. Qureshi, R.L. Buckner, B. Fischl, and A.J. van der Kouwe, "Prospective motion correction with volumetric navigators (vNavs) reduces the bias and variance in brain morphometry induced by subject motion," Neuroimage, 127 (2016) 11-22.

10
The Evolving View of Astrocytes

1. Minta, A., Kao, J. P. & Tsien, R. Y. Fluorescent indicators for cytosolic calcium based on rhodamine and fluorescein chromophores. The Journal of biological chemistry 264, 8171-8178 (1989).
2. Cornell-Bell, A. H., Finkbeiner, S. M., Cooper, M. S. & Smith, S. J. Glutamate induces calcium waves in cultured astrocytes: long-range glial signaling. Science 247, 470-473 (1990).
3. Araque, A., Parpura, V., Sanzgiri, R. P. & Haydon, P. G. Tripartite synapses: glia, the unacknowledged partner. Trends in neurosciences 22, 208-215 (1999).
4. Haydon, P. G. GLIA: listening and talking to the synapse. Nature reviews. Neuroscience 2, 185-193, doi:10.1038/35058528 (2001).
5. Halassa, M. M. & Haydon, P. G. Integrated brain circuits: astrocytic networks modulate neuronal activity and behavior. Annual review of physiology 72, 335-355, doi:10.1146/annurev-physiol-021909-135843 (2010).
6. Araque, A. et al. Gliotransmitters travel in time and space. Neuron 81, 728-739, doi:10.1016/j.neuron.2014.02.007 (2014).
7. Halassa, M. M., Fellin, T. & Haydon, P. G. The tripartite synapse: roles for gliotransmission in health and disease. Trends in molecular medicine 13, 54-63, doi:10.1016/j.molmed.2006.12.005 (2007).
8. Parpura, V. et al. Glutamate-mediated astrocyte-neuron signalling. Nature 369, 744-747, doi:10.1038/369744a0 (1994).
9. Araque, A., Li, N., Doyle, R. T. & Haydon, P. G. SNARE protein-dependent glutamate release from astrocytes. The Journal of neuroscience : the official journal of the Society for Neuroscience 20, 666-673 (2000).
10. Araque, A., Parpura, V., Sanzgiri, R. P. & Haydon, P. G. Glutamate-dependent astrocyte modulation of synaptic transmission between cultured hippocampal neurons. The European journal of neuroscience 10, 2129-2142 (1998).
11. Araque, A., Sanzgiri, R. P., Parpura, V. & Haydon, P. G. Calcium elevation in astrocytes causes an NMDA receptor-dependent increase in the frequency of miniature synaptic currents in cultured hippocampal neurons. The Journal of neuroscience : the official journal of the Society for Neuroscience 18, 6822-6829 (1998).
12. Pasti, L., Volterra, A., Pozzan, T. & Carmignoto, G. Intracellular calcium oscillations in astrocytes: a highly plastic, bidirectional form of communication between neurons and astrocytes in situ. The Journal of neuroscience : the official journal of the Society for Neuroscience 17, 7817-7830 (1997).

13. Fellin, T. et al. Neuronal synchrony mediated by astrocytic glutamate through activation of extrasynaptic NMDA receptors. Neuron 43, 729-743, doi:10.1016/j.neuron.2004.08.011 (2004).

14. Bezzi, P. et al. Prostaglandins stimulate calcium-dependent glutamate release in astrocytes. Nature 391, 281-285, doi:10.1038/34651 (1998).

15. Pascual, O. et al. Astrocytic purinergic signaling coordinates synaptic networks. Science 310, 113-116, doi:10.1126/science.1116916 (2005).

16. Halassa, M. M. et al. Astrocytic modulation of sleep homeostasis and cognitive consequences of sleep loss. Neuron 61, 213-219, doi:10.1016/j.neuron.2008.11.024 (2009).

17. Poskanzer, K. E. & Yuste, R. Astrocytes regulate cortical state switching in vivo. Proceedings of the National Academy of Sciences of the United States of America, doi:10.1073/pnas.1520759113 (2016).

18. Blanco-Suarez, E., Caldwell, A. L. & Allen, N. J. Role of astrocyte-synapse interactions in CNS disorders. The Journal of physiology, doi:10.1113/JP270988 (2016).

19. Chung, W. S., Allen, N. J. & Eroglu, C. Astrocytes Control Synapse Formation, Function, and Elimination. Cold Spring Harbor perspectives in biology 7, a020370, doi:10.1101/cshperspect.a020370 (2015).

20. Andresen, L. et al. Gabapentin attenuates hyperexcitability in the freeze-lesion model of developmental cortical malformation. Neurobiology of disease 71, 305-316, doi:10.1016/j.nbd.2014.08.022 (2014).

21. Pellerin, L. & Magistretti, P. J. Sweet sixteen for ANLS. Journal of cerebral blood flow and metabolism : official journal of the International Society of Cerebral Blood Flow and Metabolism 32, 1152-1166, doi:10.1038/jcbfm.2011.149 (2012).

22. MacVicar, B. A. & Newman, E. A. Astrocyte regulation of blood flow in the brain. Cold Spring Harbor perspectives in biology 7, doi:10.1101/cshperspect.a020388 (2015).

23. Messing, A., LaPash Daniels, C. M. & Hagemann, T. L. Strategies for treatment in Alexander disease. Neurotherapeutics : the journal of the American Society for Experimental NeuroTherapeutics 7, 507-515, doi:10.1016/j.nurt.2010.05.013 (2010).

24. Rimmele, T. S. & Rosenberg, P. A. GLT-1: The elusive presynaptic glutamate transporter. Neurochemistry international 98, 19-28, doi:10.1016/j.neuint.2016.04.010 (2016).

25. Eid, T. et al. Loss of glutamine synthetase in the human epileptogenic hippocampus: possible mechanism for raised extracellular glutamate in mesial temporal lobe epilepsy. Lancet 363, 28-37 (2004).

26. Ortinski, P. I. et al. Selective induction of astrocytic gliosis generates deficits in neuronal inhibition. Nature neuroscience 13, 584-591, doi:10.1038/nn.2535 (2010).

27. Chung, W. S. et al. Novel allele-dependent role for APOE in controlling the rate of synapse pruning by astrocytes. Proceedings of the National Academy of Sciences of the United States of America 113, 10186-10191, doi:10.1073/pnas.1609896113 (2016).

11
Understanding the Terrorist Mind

1. Tajfel, H., & Turner, J. C. (1979). An integrative theory of intergroup conflict. The social psychology of intergroup relations, 33(47), 74.
2. Cunningham, W. A., Johnson, M. K., Raye, C. L., Chris Gatenby, J., Gore, J. C., & Banaji, M. R. (2004). Separable neural components in the processing of black and white faces. Psychol Sci, 15(12), 806-813.
3. Van Bavel, J. J., Packer, D. J., & Cunningham, W. A. (2008). The neural substrates of in-group bias: a functional magnetic resonance imaging investigation. Psychol Sci, 19(11), 1131-1139.
4. Van Bavel, J. J., Packer, D. J., & Cunningham, W. A. (2008). The neural substrates of in-group bias: a functional magnetic resonance imaging investigation. Psychol Sci, 19(11), 1131-1139.
5. Hughes, B. L., Ambady, N., & Zaki, J. (2016). Trusting outgroup, but not ingroup members, requires control: Neural and behavioral evidence. Social Cognitive and Affective Neuroscience, nsw139.
6. ReChoi, J. K., & Bowles, S. (2007). The coevolution of parochial altruism and war. Science, 318(5850), 636-640.
7. Noor, M., Shnabel, N., Halabi, S., & Nadler, A. (2012). When suffering begets suffering the psychology of competitive victimhood between adversarial groups in violent conflicts. Personality and Social Psychology Review, 16(4), 351-374.
8. Pronin, E., Gilovich, T., & Ross, L. (2004). Objectivity in the eye of the beholder: divergent perceptions of bias in self versus others. Psychol Rev, 111(3), 781-799.
9. Batson, C., Polycarpou, M., Harmon-Jones, E., Imhoff, H., Mitchener, E., Bednar, L., et al. (1997). Empathy and attitudes: Can feeling for a member of a stigmatized group improve feelings toward the group? Journal of Personality and Social Psychology, 72, 105-118.
10. Davis, M. H. (1983). Measuring individual differences in empathy: Evidence for a multidimensional approach. Journal of Personality and Social Psychology, 44(1), 113-126.
11. Argo, N. (2009). Why Fight?: Examining Self-Interested Versus Communally-Oriented Motivations in Palestinian Resistance and Rebellion. Security Studies, 18(4), 651-680.
12. Kteily, N., Bruneau, E., Waytz, A., & Cotterill, S. (2015). The ascent of man: Theoretical and empirical evidence for blatant dehumanization. Journal of Personality and Social Psychology, 109(5), 901.
13. Bruneau, E., Kteily, N., & Laustsen, L. (in review). Blatant dehumanization and the European 'Refugee Crisis'.
14. Kteily, N., Hodson, G., & Bruneau, E. (2016). They see us as less than human: Metadehumanization predicts intergroup conflict via reciprocal dehumanization. Journal of Personality and Social Psychology, 110(3), 343.
15. Kteily, N., & Bruneau, E. (In Press). Backlash: The Politics and Real-World Consequences of Minority Group Dehumanization. Personality and Social Psychology Bulletin.
16. Bruneau, E., Kteily, N., & Falk, E. A hypocrisy intervention reduces collective blame and eases intergroup hostility. (in review)

12
Finding the Hurt in Pain

1. The Story of Pain: from prayer to painkillers. Joanna Bourke. Oxford University Press. 2014.
2. The Body in Pain: The making and unmaking of the world. Elaine Scarry. 1985 Oxford University Press.
3. The Challenge of Pain. Patrick Wall, Ronald Melzack. Penguin. 1996.
4. Pain: The Science of Suffering. Columbia University Press. 2000.
5. Regarding the Pain of Others. Susan Sontag. Penguin Books. 2004
6. Lee MC, Tracey I. Imaging pain: a potent means for investigating pain mechanisms in patients. Br J Anaesth. 2013 Jul;111(1):64-72.
7. Lee M, Tracey I. Neuro-genetics of persistent pain. Curr Opin Neurobiol. 2013 Feb;23(1):127-32. doi: 10.1016/j.conb.2012.11.007. Review. PubMed PMID: 23228429.
8. Tracey I. Can neuroimaging studies identify pain endophenotypes in humans? Nat Rev Neurol. 2011 Mar;7(3):173-81.
9. Institute of Medicine Relieving Pain in America,2011; www.painineurope.com
10. von Hehn CA, Baron R, Woolf CJ. Deconstructing the neuropathic pain phenotype to reveal neural mechanisms. Neuron. 2012 Feb 23;73(4):638-52.
11. Tracey I, Mantyh PW. The cerebral signature for pain perception and its modulation. Neuron. 2007 Aug 2;55(3):377-91. Review.
12. *Ploghaus, A., *Tracey, I., Gati, J.S., Clare, S., Menon, R.S., Matthews, P.M., Rawlins, J.N.P. Dissociating pain from its anticipation in the human brain (1999) Science, 284 (5422), pp. 1979-1981. *joint corresponding
13. Denk F, McMahon SB, Tracey I. Pain vulnerability: a neurobiological perspective. Nat Neurosci. 2014 Feb;17(2):192-200.
14. Berna, C., Leknes, S., Holmes, E.A., Edwards, R.R., Goodwin, G.M., Tracey, I. Induction of Depressed Mood Disrupts Emotion Regulation Neurocircuitry and Enhances Pain Unpleasantness (2010) Biological Psychiatry, 67 (11), pp. 1083-1090.
15. Wiech K, Tracey I. The influence of negative emotions on pain: behavioral effects and neural mechanisms. Neuroimage. 2009 Sep;47(3):987-94.
16. Wiech K, Ploner M, Tracey I. Neurocognitive aspects of pain perception. Trends Cogn Sci. 2008 Aug;12(8):306-13.
17. Bingel U, Tracey I. Imaging CNS modulation of pain in humans. Physiology (Bethesda). 2008 Dec;23:371-80.
18. Tracey I, Dickenson A. SnapShot: Pain perception. Cell. 2012 Mar 16;148(6):1308-1308.e2.
19. The Placebo Effect in Clinical Practice. Walter A Brown. Oxford University Press. 2013
20. Tracey I. Getting the pain you expect: mechanisms of placebo, nocebo and reappraisal effects in humans. Nat Med. 2010 Nov;16(11):1277-83.
21. Eippert F, Bingel U, Schoell ED, Yacubian J, Klinger R, Lorenz J, Büchel C. Activation of the opioidergic descending pain control system underlies placebo analgesia. Neuron. 2009 Aug 27;63(4):533-43.
22. Bingel, U., Wanigasekera, V., Wiech, K., Mhuircheartaigh, R.N., Lee, M.C., Ploner, M., Tracey, I. The effect of treatment expectation on drug efficacy: Imag-

ing the analgesic benefit of the opioid remifentanil (2011) Science Translational Medicine, 3 (70), art. no. 70ra14.

23. Leknes S, Tracey I. A common neurobiology for pain and pleasure. Nat Rev Neurosci. 2008 Apr;9(4):314-20.

24. Leknes, S., Berna, C., Lee, M.C., Snyder, G.D., Biele, G., Tracey, I. The importance of context: When relative relief renders pain pleasant (2013) Pain, 154 (3), pp. 402-410.

25. Wager TD, Atlas LY, Lindquist MA, Roy M, Woo CW, Kross E. An fMRI-based neurologic signature of physical pain. N Engl J Med. 2013 Apr 11;368(15):1388-97.

26. Duff EP, Vennart W, Wise RG, Howard MA, Harris RE, Lee M, Wartolowska K, Wanigasekera V, Wilson FJ, Whitlock M, Tracey I, Woolrich MW, Smith SM. Learning to identify CNS drug action and efficacy using multistudy fMRI data. (2015) Science Translational Medicine, 7(274):274ra16.

Index

for Alzheimer's disease, 21
biochemical mechanisms of, 19–20
as brain protector, 16, 17–19
chemistry of, 16
on CREB, 20
for Fragile X syndrome, 22
history of, 17
on inflammation, 21–22
limitations of, 23
medical uses of, 16
in nature, 16
potential uses of, 20–22
on suicide with bipolar disorder, 22
LSD-like NPS, 4

M

Magnetic resonance imaging (MRI),
96–97. *See also* Human
Connectome Project;
Neuroimaging
diffusion, 97
resting-state fMRI, 97
structural, 96
task-activated functional, 97
Malignancy, infectious, 26–27
Malignant protein, 25–35
Aβ amyloid in, 31–32
in infectious cancers, 26–27
in infectious neurodegenerative
diseases, 26–27
in plaques and tangles, 31–32
prion infectivity and, 33–35
prion paradigm in, 34–35
prions as, 27–29, 27f
in scrapie, 28
seeding in, 27
Manganese
on DNA methylation, 75
in drinking water, 77–78
in water and food, 76
Manji, Husseini, 17
Marijuana, *Stoned: A Doctor's Case for
Medical Marijuana* (Casarett),
151–156
Marijuana-like NPS, 4, 8f, 10–12
Marquez, Patricio V., 83
McKhann, Guy M., ix

4-MEC, 6–9, 7f
Memento, 145
Memory
amnesia and
anterograde, 145
in film, 147–148
electro-convulsive therapy on, 148
formation of, 39, 44–45
interrogation and torture on, 159
reconstructive, 159
recovered, 148
Memory & Movies (Seamon), 145–149
Mental health, as global priority, 83–95
collaboration and financing in,
91–93
economic burden of, 85
economic loss and return on
investment in, 86–87
epidemiology of, 84–85
for migrants and refugees, 88–89
neuroscience relevance in, 90–91
parity in, 87–88
risk factors in, 84
social costs in, 84
technological solutions in, 89
view and treatment of, 85
in workplace, 90
in World Bank Group/World
Health Organization
Global Mental Health
Event, 85–86
Mephedrone, 6–9, 7f
Mercury, on DNA methylation, 75
Metabolism, brain, astrocyte support
of, 112
Methylenedioxypyrovalerone (MDPV),
6–9, 7f
Methylmercury, 77–78
Methylone, 6–9, 7f
Migrants, mental health for, 88–89
"Molly," 6–9, 7f, 13
Monoamine transporters, drugs acting
at, 7–8
Multimodal Surface Matching (MSM)
algorithm, 102
Multiple sclerosis, relapsing-remitting,
beta-interferons for, 59–60